T0363788

Submarines

300 of the World's Greatest Submersibles

Robert Jackson

amber
BOOKS

Published by
Amber Books Ltd
United House
North Road
London
N7 9DP
United Kingdom
www.amberbooks.co.uk
Facebook: amberbooks
YouTube: amberbooksltd
Instagram: amberbooksltd
X(Twitter): @amberbooks

ISBN: 978-1-83886-515-3

Additional text: Charles Catton
Design: Hawes Design

Printed in China

Picture Credits:
AirSeaLand Photos

Artwork Credits
All artworks Istituto Geografico De Agostini S.p.A. except the following:
Amber Books 62, 216, 224, 241
Amber Books/Patrick Mulrey 24, 215, 217, 242, 243, 274
BAe Systems 25
Mainline Design (Guy Smith) 17, 171, 177, 188, 214, 219, 225, 237, 271, 291, 293
Tony Gibbons 16, 30, 40, 44, 99, 102, 120, 141, 168, 172, 181, 189, 190,
191, 192, 195, 201, 202, 206

Submarines

300 of the World's
Greatest Submersibles

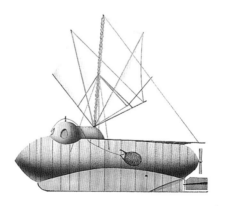

CONTENTS

Introduction 7

A1 to Astute 14

B1 to Bronzo 26

C1 to Corallo 35

D1 to Dykkeren 47

E11 to Explorer 70

F1 to Fulton 86

G1 to Gymnôte 104

H1 to Hvalen 135

I7 to Kilo 148

L3 to Marsopa 161

N1 to Nymphe 170

O class to Porpoise 182

R1 to Rubis 197

S1 to Swordfish 210

Tang to Typhoon 231

U1 to Upholder 245

V class to Whiskey 268

X1 to Zoea 281

Underwater Weapons 291

Index 313

Introduction

The submarine has revolutionized naval warfare. These vessels, which wage war beneath the waves, have progressed from the crude, steam-driven craft of the American Civil War to silent nuclear submarines that can cruise for months underwater without surfacing, and which carry intercontinental missiles mounting multiple nuclear warheads.

The concept of submarine warfare is centuries old. In 1634, two French priests, Fathers Mersenne and Fornier, produced quite a detailed design for an armed underwater craft, and in 1648, John Wilkins, Oliver Cromwell's brother-in-law, discussed the possibilities of a 'Submarine Ark'. It was the American War of Independence that saw the first steps in the evolution of the submarine as a true fighting weapon. The first ever underwater was mission was undertaken in September 1776, when an American soldier, Ezra Lee, operating a small submersible called *Turtle*, tried to attach an explosive charge to HMS *Eagle* in the Hudson River. Although not a true submersible, all of Turtle was under the surface when she was in action except for a tiny conning tower fitted with glass ports so that the sole occupant could find his way to the target. Ezra Lee failed in his mission to attach an explosive charge to the hull of the enemy ship, and the little craft was finally lost when the frigate that was transporting her ran aground.

The Napoleonic Wars might have seen the use of submersible craft in some numbers, had the ideas of an American inventor called Robert Fulton

Above: HMS M2, was launched in October 1918, and thus saw no war service. She became a seaplane carrier in April 1928, and foundered off Portland in 1932.

met with success. Having failed to arouse interest with his submarine project in America, Fulton went to France in 1797, where the plans for his prototype submarine were accepted. Launched under the name *Nautilus* in 1800, she was to become the first submarine to be built to a government contract. During trials in Le Havre harbour, *Nautilus* remained underwater at a depth of 7.6m (25ft) for one hour. After the French lost interest in the project Fulton took his design to Britain, where he persuaded Prime Minister William Pitt to examine the idea, but she was not adopted. The attitude of the British Admiralty to the concept was summed up by Lord St Vincent, who denounced Pitt as, 'the greatest fool that ever existed to encourage a mode of warfare which those who commanded the sea did not want, and which, if successful, would deprive them of it.'

In the long peace that followed the defeat of Napoleon there was little incentive for inventive minds to pursue the development of submarine craft, and it took the onset of the American Civil War to create a fresh upsurge of interest. The designs that emerged, however, were little more than suicide craft, armed with an explosive charge on the end of a long pole – the so-called 'spar torpedo'. *H L Hunley*, named after her inventor was the first true submersible craft to be used successfully against an enemy. On 17 February 1864, she sank the Union ship *Housatonic*, but was dragged down with her by the wave created by her spar torpedo. Years later, the wreck was located on the sea bed, the skeletons of eight of her crew still seated at their crankshaft.

THE 'DAMNED, UN-ENGLISH' WEAPON

It was the Americans who took the lead in submarine design and development as the turn of the century drew near, and at the forefront was an Irish-American called John P. Holland. Holland's first successful submarine design was his *No 1*; the diminutive craft was originally designed to be hand-cranked like previous submarines, but with the introduction of the newly-developed Brayton four-horsepower petrol engine, Holland was able to produce a more reliable vessel. *Holland No 1* was built at the Albany Iron Works and was completed in 1878. She is now housed in the Paterson Museum, USA. Holland's far-sighted faith in the petrol engine proved premature, other submarine designs of this period being still dependent on steam for their motive power.

The first American submarine of what might be called 'modern' design was *Holland VI*, which later became the prototype for British and Japanese submarines, which combined petrol engine and battery power with hydroplanes. *Holland VI* entered service with the US Navy as the *Holland* in 1900. Although the American press praised the little Holland submarine and bestowed such lurid descriptions as 'Monster War Fish' on her, she was in fact a very primitive craft.

Admiral Sir Arthur Wilson's irascible speech in 1899 denouncing submarines as, 'underhand, unfair and damned un-English' has often been quoted to suggest that the Admiralty had closed its mind on the subject of submarines. The Royal Navy had been alarmed by the sudden proliferation of submarines in the French and American fleets, and in fact an internal study was already underway. The 1901-2 Naval Estimates made provision for the building of five improved boats of the Holland type (an American design) for evaluation. The first five boats to be commissioned were built under licence by Vickers at Barrow-in-Furness; the company and the Navy's newly-appointed Inspecting-Captain of Submarines, Captain Reginald Bacon, set about making a series of improvements, so that when HM Submarine *No 1* was launched on 2 November 1902 she bore little resemblance to her American progenitor. In March 1904 all five boats of the A class, as they were now called, took part in a simulated attack on the cruiser *Juno* off Portsmouth. It was successful, but *A1* was involved in a collision with a passenger liner and sank with the loss of all hands. In all, 13 A-class boats were built, followed by 11 B class and 38 C-class. From now on, the submarine was to be one of the Royal Navy's principal weapons of war.

Above: Cramped living conditions inside HMS Graph, *formerly U-boat* 570.

By 1914 the Americans, British, French, Italians and Russians all had substantial submarine fleets. The Germans were slow at first to catch up, but as World War I progressed the German Navy's submarine service became increasingly competent and its boats technologically more advanced, until by the end of 1916 it had become the navy's main offensive arm. Germany's large, long-range 'cruiser' U-boats were a revelation to the Allies, and very nearly brought them to their knees.

In 1917 the threat to British merchant ships – which all sailed independently – from U-boats was dire, and in April that year they sank 907,000 tonnes (893,000 tons) of shipping, of which 564,019 tonnes (555,110 tons) were British vessels. It was only the belated introduction of the convoy system that turned things round for the Allies.

The British developed other countermeasures. They built the so-called Dover Barrage, consisting of heavily-armed ships moored in lines across the Channel with minefields, nets and other obstacles between them. However, the vast expenditure of time, labour and materials that the Barrage involved yielded few fruits, with only four U-boats being sunk by the end of 1917. During 1918, (by which time the U-boats had already been defeated in the Atlantic by the convoy system), the barrage did claim between 14 and 26 submarines.

WORLD WAR II: THE BATTLE FOR THE OCEANS

The early successes of the German Navy's U-boat arm were not forgotten in the inter-war years by Nazi Germany's naval planners. The ocean-going U-boat was to prove a fearsome weapon in World War II, and one that came close to bringing Britain to her knees in the bitter conflict known as the Battle of the Atlantic. Admiral Karl Dönitz developed the 'wolf-pack' tactic in an attempt to counter the effectiveness of the convoy system which had defeated him 20 years before. Despite the technical excellence of their boats and the skill and courage of their crews, the Germans lost that battle because the Allies, in partiular the British, gradually gave air cover to their convoys, fear of which forced the U-boats to dive; they were thus incapable of keeping up with their targets at their slower submerged speeds.

Long-range maritime patrol aircraft, equipped with new detection radar, enjoyed increasing success in locating and destroying the U-boats around their convoys. The same aircraft co-operated on a magnificent scale with Royal Navy and US Navy hunter-killer groups, composed of fast destroyers and frigates that combed the ocean for submarines in a highly-organized pattern and hounded them to destruction in the last two years of the war.

The final cost to the German Navy's U-boat Service was appalling. Of the 1162 U-boats built and commissioned during the war, 785 were recorded as 'spurlos versenkt' (lost without trace). Of the 40,000 officers and men who served in U-boats from 1939 to 1945, 30,000 never returned. Yet there was no escaping the fact that the U-boats, in those years of war, had sunk 2828 Allied merchant ships totalling 14,923,052 tonnes (14,687,321 tons). British naval casualties alone were 80,000, of which over 30,000 were merchant seamen.

The Battle of the Atlantic ended in victory for the Allies; but on the other side of the world, in the Pacific, long-range American submarines succeeded where the German U-boats had failed. From 1942, in an increasingly effective campaign against the Japanese merchant fleet – which was not put into convoy – they sank 5,588,275 tonnes (5,500,000 tons) of shipping for the loss of 49 submarines, and in the process they brought the island nation of Japan to a virtual standstill.

The fact that the exploits of the American submariners remained a backwater of history until long after the war was due partly to inter-service rivalries, but in the main to the need to preserve one of the war's most closely-guarded secrets: the fact that the Allies, because of high-grade 'Ultra' intelligence, had access to the enemy's naval codes and therefore knew a great deal about the movements of his maritime traffic. In the Atlantic, this intelligence contributed to the safe routing of convoys. With hundreds of ships concentrated into just a few convoys, the enemy had far fewer targets which were thus far harder to find. Of all Atlantic convoys sailed across the Atlantic during the conflict, ninety per cent of them never sighted a U-boat.

World War II saw enormous strides in the development of underwater weapons such as acoustic homing torpedoes, anti-submarine rockets and

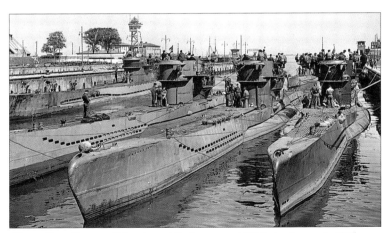

Above: Four German U-boats and their crews at Wilhelmshaven in 1945, prior to being handed over to the Allies.

depth charges. A section on the modern descendants of these weapons will be found at the end of this book.

RUN SILENT, RUN DEEP: THE COLD WAR ERA

Ballistic-missile Submarines: The concept of the missile submarine is not new, dating as it does to German plans of World War II. But it was not until the 1950s that the Americans and Russians began to explore the concept of the nuclear-powered ballistic-missile submarine, a vessel capable of remaining submerged for lengthy periods, making use of the polar ice-cap and various oceanic features to remain undetected. Armed with nuclear-tipped rockets, it would be the ultimate deterrent.

Although the Americans were the first to make the nuclear-powered submarine breakthrough, with an early class of boat based on the prototype *Nautilus*, what the US Navy really wanted was to merge the new technologies of ballistic missiles, smaller thermonuclear weapons, inertial guidance systems and nuclear weapons into a single weapon system. They succeeded with the deployment, in 1960, of the first Fleet Ballistic-Missile (FBM) submarine, armed with the Polaris A1 missile.

The Russians were quick to respond, deploying the Hotel-class nuclear ballistic-missile submarine. This was armed initially with three SSN-4 Sark missiles, with a range of only 350 nautical miles, (650km), but after 1963 it converted to the SSN-5 Serb, with a range of 650nm (1200km). The ballistic-missile submarine race was on, and it would later be joined by Britain, France and China. By the 1980s, the missile submarine (SSBN) had become an awesome weapon of destruction, capable of carrying up to sixteen missiles, each with multiple warheads, that could rain nuclear destruction on targets 2,500nm (4600km) from their launch point.

Nuclear-attack Submarines: For nearly three decades, NATO and Warsaw Pact submariners played a potentially deadly game of cat and mouse in the depths of the world's

Above: SSBN Resolution, *launched in 1966, which carried the Polaris missile.*

oceans – the tools of their trade being nuclear attack and hunter-killer submarines (SSNs), packed with weaponry and sensors. Their targets were the ballistic-missile submarines and naval task forces of the enemy.

The development of nuclear-attack submarines in the United States began at about the same time, but the designs followed different paths. The Americans concentrated on anti-submarine warfare (ASW) and the Russians on a multi-mission role, encompassing both ASW and surface attack with large anti-ship cruise missiles. Later on the Americans also adopted a multi-mission capability with the deployment of submarine-launched weapons like Sub-Harpoon and Tomahawk, designed for anti-ship and land attack.

The main advantages of the nuclear-attack submarine are its ability to remain submerged for virtually unlimited periods, its deep-diving capability, the sophisticated long-range sensor systems that it carries, and the high power output of its reactor that can be converted into very high underwater speeds. The later generations of nuclear-attack submarines are virtually under-water cruisers; the Russian Oscar class, for example, was the underwater equivalent of the Kirov class of battlecruiser. Their combat arena, in the main, lay under the Arctic ice cap, once considered to be a safe haven for the ballistic-missile submarines.

Diesel-electric Submarines: In some naval circles, it was thought that the advent of the nuclear submarine would mean the demise of the old-fashioned diesel-powered boats, a form of propulsion which had carried the submarine through all the stages of its development since World War I. This was not the case.

Only the richest nations can afford the costly nuclear powerplants that are necessary for long-endurance ocean patrol; for other countries, the diesel-electric boat provides a cost-effective solution to the problem of maintaining an undersea presence in territorial waters. Diesel-electric boats also have a considerable advantage in that they are very quiet when running on their electric motors underwater, which makes them very hard to detect. During the 1982 Falklands War, the Royal Navy failed to locate an Argentine Navy Type 209 submarine, the *San Luis*, which made three abortive attacks on the British task force.

Today navies long dedicated to hunting submarines in the ocean depths and now occupied threats to commercial shipping, often in restricted waters like the Gulf and Adriatic, must contend with the definite threat of the small diesel submarine in potentially hostile hands.

A1

The A-class vessels were the first submarines designed in Britain, although they were originally based on the earlier US Holland type which had entered Royal Navy service in 1901. A1 was basically a slightly lengthened *Holland*, but from A2 onwards they were much larger. They were also the first submarines to be made with a proper conning tower to allow surface running in heavy seas. Originally fitted with a single bow-mounted torpedo tube, the class was equipped with a second one from A5 onwards. Built by Vickers, the class helped the Royal Navy to develop and refine its submarine doctrine and operating skills. Thirteen boats were built between 1902 and 1905, and some served in a training capacity during World War I. One of the class, A7, was lost with all hands when she dived into the sands at Whitesand Bay.

Country:	Britain
Launch date:	July 1902
Crew:	11
Displacement:	Surfaced: 194 tonnes (191 tons)
	Submerged: 274.5 tonnes (270 tons)
Dimensions:	30.5m x 3.5m (100ft x 10ft 2in)
Armament:	Two 460mm (18in) torpedo tubes
Powerplant:	One 160hp petrol engine, one 126hp electric motor
Surface range:	593km (320nm) at 10 knots
Performance:	Surfaced: 9.5 knots
	Submerged: 6 knots

Acciaio

The *Acciaio* was lead vessel of a class of 13 submarines built in 1941-42. Nine were lost during World War II, including *Acciaio* herself, torpedoed and sunk by HM submarine *Unruly* north of the Messina Straits on 13 July 1943. The longest surviving vessel of the class was *Giada*, which was removed from the naval list in February 1948 under the terms of the Peace Treaty and converted to a hull for recharging batteries. She reappeared on the naval list in March 1951, and was rebuilt and modified to carry four 533mm (21in) torpedo tubes forward; no gun armament was fitted. She was definitively discarded in January 1966. Another boat, *Nichelio*, was transferred to Soviet Russia under the Peace Treaty in February 1949 and was designated *Z14*; she was scrapped about 1960. Some boats of the class were powered by different engines.

Country:	Italy
Launch date:	20 July 1941
Crew:	46-50
Displacement:	Surfaced: 726 tonnes (715 tons) Submerged: 884 tonnes (870 tons)
Dimensions:	60m x 6.5m x 4.5m (197ft x 21ft 4in x 14ft 9in)
Armament:	Six 533mm (21in) torpedo tubes; one 100mm (3.9in) gun
Powerplant:	Two diesels, two electric motors
Surface range:	7042km (3800nm) at 10 knots
Performance:	Surfaced: 15 knots Submerged: 7.7 knots

Agosta

Designed by the French Directorate of Naval Construction as very quiet but high-performance ocean-going diesel-electric boats (SSKs), the Agosta-class boats were each armed with four bow torpedo tubes equipped with a rapid reload pneumatic ramming system that could launch weapons with a minimum of noise signature. The tubes were of a completely new design when the Agostas were authorized in the mid-1970s, allowing a submarine to fire its weapons at all speeds and at any depth down to its maximum operational limit, 350m (1148ft). Two of the four boats, *Agosta* and *Beveziers,* were paid off in the early 1990s; the two remaining, *La Praya* and *Ouessant,* were based at Brest from June 1995 and assigned to the Atlantic Attack Submarine Group. *La Praya* was decommissioned in 2001; *Ouessant* a year later. She is now a museum ship in Kiebang, Malacca.

Country:	France
Launch date:	19 October 1974
Crew:	54
Displacement:	Surfaced: 1514 tonnes (1490 tons) Submerged: 1768 tonnes (1740 tons)
Dimensions:	67.6m x 6.8m x 5.4m (221ft 9in x 22ft 4 in x 17ft 9in)
Armament:	Four 550mm (21.7in) torpedo tubes; 40 mines
Powerplant:	Two diesels, one electric motor
Surface range:	15,750km (8500nm) at 9 knots
Performance:	Surfaced: 12.5 knots Submerged:17.5 knots

Albacore

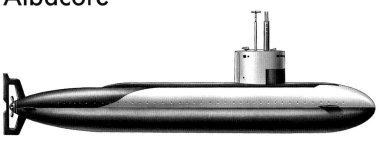

The USS *Albacore* was a high-speed experimental submarine (AGSS), conventionally powered but of radical design, with a new hull form that made her faster and more manoeuvrable than any other conventional submarine. Officially described as a hydrodynamic test vehicle, she was highly streamlined; her hull was whaleshaped, without a flat-topped deck, and her conning tower resembled a fish's dorsal fin. *Albacore* underwent several conversions during her test career. In 1959 she was fitted with an improved sonar system, an enlarged dorsal rudder, and dive brakes on the after sail section; in 1961 she received contra-rotating electrical motors, with two propellers contra-rotating about the same axis; and in 1962 she was equipped with a high capacity long-endurance silver zinc battery.

Country:	USA
Launch date:	1 August 1953
Crew:	52
Displacement:	Surfaced: 1524 tonnes (1500 tons)
	Submerged: 1880 tonnes (1850 tons)
Dimensions:	62.2m x 8.4m x 5.6m (204ft x 8ft 5in x 15ft 7in)
Armament:	None
Powerplant:	Two diesels, one electric motor
Surface range:	Not released
Performance:	Surfaced: 25 knots
	Submerged: 33 knots

Alfa

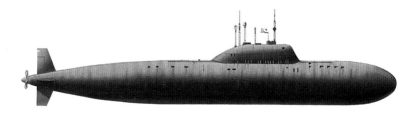

The second Russian titanium-hulled submarine design, the Project 705 Lira, known in the west as Alfa, came to light in December 1971, when the first unit was commissioned. Five more followed in 1972–82. A single reactor and turbine plant drive the boat at a phenomenal 42 knots under water. When British and American submariners first encountered Alfa they were astounded, but what was not realized at the time was that there was a serious flaw in the lead-bismuth system of Alfa's 40,000hp reactor cooling system. The plant was very unreliable, and the cost led to the *Lira/Alfa* being nicknamed the 'Golden Fish'. In addition, the design was not stressed for deep diving, as was assumed in the West – with the result that NATO navies allocated massive R&D funding to the development of deep-running torpedoes.

Country:	Russia
Launch date:	1970
Crew:	31
Displacement:	Surfaced: 2845 tonnes (2800 tons) Submerged: 3739 tonnes (3680 tons)
Dimensions:	81m x 9.5m x 8m (265ft 9in x 31ft 2in x 26ft 3in)
Armament:	Six 533mm (21in) torpedo tubes; conventional or nuclear torpedoes; 36 mines
Powerplant:	Liquid-metal reactor, two steam turbines
Range:	Unlimited
Performance:	Surfaced: 20 knots Submerged: 42 knots

Aluminaut

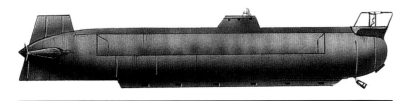

Aluminaut became famous when she was used to help recover an H-bomb which had fallen from an American B-52 bomber involved in a mid-air collision with its KC-135 tanker aircraft over Spain in 1966. Built in 1965, *Aluminaut* is capable of exploring to depths of 4475m (14,682ft). She is equipped with a side-scan sonar which builds up a map of the terrain on either side. Most routine underwater exploration never reaches such depths. Even the most advanced military submarines go down no further than 900m (2952ft): any deeper and the costs of hull-strengthening and engine up-rating are prohibtive. For commercial and scientific work in coastal waters, manned submerisbles dominate. Pressures are within comfortable engineering limits, and the crews have made impressive advances in undersea archaeology and oil exploration.

Country:	USA
Launch date:	1965
Crew:	3
Displacement:	Surfaced: not known Submerged: 81 tonnes (80 tons)
Dimensions:	not known
Armament:	none
Powerplant:	not known
Operational depth:	4475m (14,682ft)
Performance:	Surfaced: not known Submerged: not known

Aradam

A *radam* was one of the Adua class of 17 short-range vessels. These had a
double hull with blisters and were a repeat of the previous Perla class. They
gave good service during World War II, and although their surface speed was low
they were strong and very manoeuvrable. The early boats of the class took part
in the Spanish Civil War and all except one (the *Macalle,* which was in the Red
Sea) operated in the Mediterranean. Only *Alagi* survived World War II. *Aradam*
was scuttled in September 1943 in Genoa harbour to avoid capture by the
Germans, who later raised her. She was sunk by Allied bombing in the following
year. The class leader, Adua, was depth-charged and sunk by the destroyers
Gurkha and Legion on 30 September 1941. Three other units of this class, the
original *Ascianghi, Gandar* and *Neghelli,* were sold to Brazil prior to launching.

Country:	Italy
Launch date:	15 November 1936
Crew:	45
Displacement:	Surfaced: 691 tonnes (680 tons) Submerged: 880 tonnes (866 tons)
Dimensions:	60.2m x 6.5m x 4.6m (197ft 6in x 21ft 4in x 15ft)
Armament:	Six 530mm (21in) torpedo tubes, one 100mm (4in) gun
Powerplant:	Two diesel engines, two electric motors
Surface range:	4076km (2200nm) at 10 knots
Performance:	Surfaced: 14 knots Submerged: 7 knots

Archimede

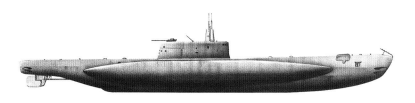

The Italian Navy operated five submarines in the Brin class, all completed in 1938-39. The last two, *Archimede* and *Torricelli,* were built in secret to replace two submarines of the same name which had been transferred to the Nationalist forces during the Spanish Civil War. The Brin class were efficient, streamlined vessels, and had a very long range. They had a partial double hull, with four torpedo tubes in the bow and four in the stern. As first completed, the submarines had two 100mm (3.9in) guns. At the outbreak of World War II *Archimede* was operating in the Red Sea and the Indian Ocean, where she remained until May 1941. She then made an epic journey round the Cape of Good Hope to Bordeaux, from where she began operations in the Atlantic. She was sunk by Allied aircraft off the coast of Brazil on 14 April 1943.

Country:	Italy
Launch date:	5 March 1939
Crew:	58
Displacement:	Surfaced: 1032 tonnes (1016 tons)
	Submerged: 1286 tonnes (1266 tons)
Dimensions:	72.4m x 6.7m x 4.5m (237ft 6in x 22ft x 15ft)
Armament:	Eight 533mm (21in) torpedo tubes, one 100mm (4in) gun
Powerplant:	Two diesels, two electric motors
Surface range:	18,530km (10,000nm) at 10 knots
Performance:	Surfaced: 17 knots
	Submerged: 8 knots

Argonaut

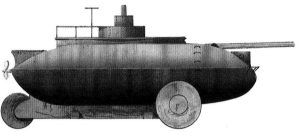

A*rgonaut* was built by Simon Lake at his own expense as a salvage vessel for inshore waters. A 30hp gasoline engine drove the single screw, and the engine could be connected to the twin front wheels for movement along the sea bed; a third wheel aft steered the craft. There was an air chamber forward so that divers could enter and leave. The vessel was rebuilt in 1899 and once made a trip of 3200km (1725nm) on the surface. Successful trials led to a number of export orders, but by that time Lake had lost the initiative to John Holland in the eyes of the US Navy, whose senior officers were not impressed by the idea of a wooden-hulled craft trundling along the sea bed. But the idea was to be resurrected nearly a century later, when designs for 'ocean crawling' submarines were again proposed.

Country:	USA
Date of launch:	1897
Crew:	5
Displacement:	Surfaced: Not known Submerged: 60 tons (59 tons)
Dimensions:	11m x 2.7m (36ft x 9ft)
Armament:	None
Powerplant:	Gasoline engine
Surface range:	not known
Performance:	Surfaced: 5 knots
	Submerged: 5 knots

Argonaut

Argonaut was the only purpose-built minelaying submarine to serve in the US Navy. Her pre-war pendant number was A1; she was later designated SS166. Apart from *Argonaut*, the only minelayers in the US Navy's inventory in 1941 were two old converted coastal passenger ships. On 1 December 1941, with the threat of war with Japan looming, she was stationed with another submarine, USS *Trout,* off Midway for reconnaissance purposes. Shortly after the attack on Pearl Harbor she was converted to the transport and special duties role, and on 17 August 1942, together with the USS *Nautilus*, she landed the 2nd Raider Battalion in Makin, in the Gilbert Islands, and extracted the force after it had attacked enemy installations. On 10 January 1943, *Argonaut* failed to return from a special operation off Lae.

Country:	USA
Launch date:	10 November 1927
Crew:	89
Displacement:	Surfaced: 2753 tonnes (2710 tons) Submerged: 4145 tonnes (4080 tons)
Dimensions:	116m x 10.4m x 4.6m (381ft x 34ft x 15ft 6in)
Armament:	Four 533mm (21in) torpedo tubes, two 152mm (6in) guns, 60 mines
Powerplant:	Two-shaft diesels, electric motors
Surface range:	10,747km (5800nm) at 10 knots
Performance:	Surfaced: 15 knots Submerged: 8 knots

Arihant

India's first indigenously-built nuclear-powered ballistic missile submarine (SSBN) was commissioned in 2016. Named after the Sanskrit word meaning 'destroyer of enemies', *Arihant* signifies India's commitment to maintaining a credible nuclear deterrent. She carries 12 K-15 Sagarika missiles or 4 K-4 missiles, providing India with a second-strike capability in the event of a nuclear attack. *Arihant's* construction involved extensive collaboration between India's Defense Research and Development Organization (DRDO), the Indian Navy, and various domestic and international defence contractors. In October 2022 the vessel launched a test K-15 SLBM in the Bay of Bengal, successfully hitting the designated target area. A sister vessel, INS *Arighat*, is an improved version of the *Arihant*, is believed to be now in service, with two further larger vessels planned.

Country:	India
Launch date:	26 July 2009
Crew:	96
Displacement:	Surfaced: 6000 tonnes (5900 tons) Submerged: N/A
Dimensions:	110m x 11m x 9m (361ft x 36ft x 29ft)
Armament:	Six 533mm (21in) launch tubes; 30 torpedoes; 12 K-15 SLBM
Powerplant:	One pressurized water reactor, 111,000hp (83MW), single screw
Surface range:	Unlimited
Performance:	Surfaced: 15 knots Submerged: 24 knots

Astute

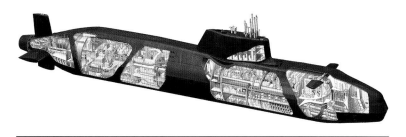

The Royal Navy's newest class of nuclear-powered attack submarines, the Astute class represent a significant leap forward in technology, replacing the older Swiftsure and Trafalgar-class submarines. Despite significant delays and cost overruns during development, they have emerged as some of the most advanced in the world, featuring state-of-the-art sonar and combat management systems. Built at Barrow-in-Furness, the first of the class, *Astute*, was commissioned in 2010 with another four – *Ambush*, *Artful*, *Audacious* and *Anson* – currently in service, and the final two vessels currently under construction, *Agamemnon* and *Agincourt*, due to be commissioned by 2024 and 2026 respectively. The Royal Navy intend to replace the Astute class with the new AUKUS class currently in development in the late 2030s.

Country:	Britain
Launch date:	8 June 2007
Crew:	98
Displacement:	Surfaced: 7000 tonnes (6889 tons) Submerged: 7400 tonnes (7283 tons)
Dimensions:	97m x 11.3m x 10m (318ft x 37ft x 33ft)
Armament:	Six 533mm (21in) launch tubes for Spearfish torpedoes/Tomahawk Block IV TLAM; 38 weapons
Powerplant:	One RR PWR2 Core H reactor, two turbines, single shaft, propulsor
Surface range:	Unlimited
Performance:	Surfaced: N/A Submerged: 29 knots

B1

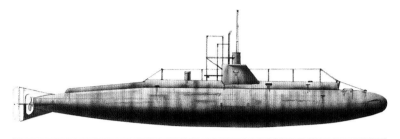

Construction of the improved B-class submarines was under way before the A-class boats were completed. An extended superstructure on top of the hull gave improved surface performance, while small hydroplanes on the conning tower improved underwater handling. By 1910 the Royal Navy had 11 B-class boats. They were not comfortable craft in which to serve; their interior stank of raw fuel, bilge-water and dampness, all pervaded by a stench of oil, and when submerged there was a constant risk of explosion from violent sparks produced by unshieded electrical components in an atmosphere saturated with petrol vapour. Six B-class submarines were sent to Gibraltar and Malta. *B1* was broken up in 1921. The first RN VC of World War I was awarded to a B-class boat commander.

Country:	Britain
Launch date:	October 1904
Crew:	16
Displacement:	Surfaced: 284 tonnes (280 tons) Submerged: 319 tonnes (314 tons)
Dimensions:	41m x 4.1m x 3m (135ft x 13ft 6in x 9ft 10in)
Armament:	Two 475mm (18in) torpedo tubes
Powerplant:	Single screw petrol engine, electric motor
Surface range:	2779km (1500nm) at 8 knots
Performance:	Surfaced: 13 knots Submerged: 7 knots

Balilla

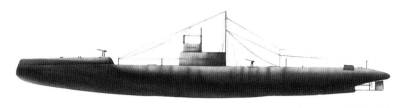

Balilla was originally ordered by the Germany Navy from an Italian yard and was allocated the number *U42*, but the boat never saw German service. Taken over by the Italian Navy in 1915 and named *Balilla,* she saw some service in the Adriatic, but while on patrol on 14 July 1916 she was sunk by Austrian torpedo boats with the loss of all 38 crew members. The principal task of the Italian submarines operating in the Adriatic in World War I was to patrol the coastline of Dalmatia, which had many harbours and inlets that were used by the Austro-Hungarian fleet. Operating conditions were difficult, as the waters were fairly shallow and it was not easy to take evasive action if attacked. Both Austrians and Italians made extensive use of seaplanes once it was found that submerged submarines could often be seen from them.

Country:	Italy
Launch date:	August 1913
Crew:	38
Displacement:	Surfaced: 740 tonnes (728 tons)
	Submerged: 890 tonnes (876 tons)
Dimensions:	65m x 6m x 4m (213ft 3in x 19ft 8in x 13ft 1in)
Armament:	Four 450mm (17.7in) torpedo tubes, two 76mm (3in) guns
Powerplant:	Two-shaft diesel/electric motors
Surface range:	7041km (3800nm) at 10 knots
Performance:	Surfaced: 14 knots
	Submerged: 9 knots

Barbarigo

The first *Barbarigo* was one of a group of four medium-sized submarines laid down in October 1915 but not completed until the end of World War I. The batteries were placed in four watertight compartments under the horizontal deck that ran the full length of the vessel. This was a new arrangement, as the batteries were usually concentrated in one large compartment for ease of access, and was designed as a safety measure to prevent the release of chlorine gas in large quantities in the event of seawater infiltrating into the boat and making accidental contact with the batteries. The Barbarigo class had a range of just over 3218km (1734nm) at 11 knots, but their maximum diving depth was only 50m (164ft). The Italians never managed to exploit their submarines to the full. *Barbarigo* was sold in 1928.

Country:	Italy
Launch date:	November 1917
Crew:	35
Displacement:	Surfaced: 774 tonnes (762 tons) Submerged: 938 tonnes (923 tons)
Dimensions:	67m x 6m x 3.8m (220ft x 19ft 8in x 12ft 6in)
Armament:	Six 450mm (17.7in) torpedo tubes, two 76mm (3in) guns
Powerplant:	Twin shaft diesel/electric motors
Surface range:	3218km (1734nm) at 11 knots
Performance:	Surfaced: 16 knots Submerged: 9.8 knots

Barbarigo

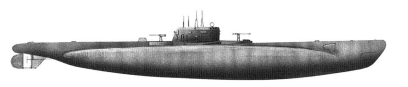

Barbarigo was one of a class of nine units, only one of which survived World War II. These vessels were built with a partial double hull and internal ballast tanks. They were reasonably fast and manoeuvrable, but had poor transverse stability. This characteristic was aggravated by the long conning tower, a common feature of Italian submarines during this period. The class also had a relatively short range, some 1425km (768nm) on the surface and 228km (123nm) submerged at three knots. Maximum diving depth was about 100m (330ft). *Barbarigo* was completed in 1939 and, after more than four years of active service, she was converted into a transport submarine to ferry supplies to Japan. Her first such voyage was in June 1943, but she was sighted by Allied aircraft on the surface in the Bay of Biscay, attacked and sunk.

Country:	Italy
Launch date:	13 June 1938
Crew:	58
Displacement:	Surfaced: 1059 tonnes (1043 tons) Submerged: 1310 tonnes (1290 tons)
Dimensions:	73m x 7m x 5m (239ft 6in x 23ft x 16ft 6in)
Armament:	Eight 533mm (21in) torpedo tubes
Powerplant:	Twin shaft diesel/electric motors
Surface range:	1425km (768nm) at 10 knots
Performance:	Surfaced: 17.4 knots Submerged: 8 knots

Bass

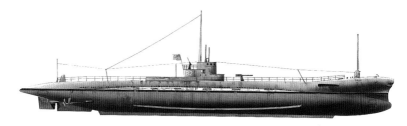

The USS *Bass* was one of three submarines of the Barracuda class, built at the Portsmouth Navy Yard in the mid-1920s. The first post-World War I submarines, these fast fleet boats were the first of nine authorized as part of the major 1916 programme. They were about twice the size of the earlier S-class boats, and were even bigger than the three wartime T-class craft broken up under the terms of the 1930 London Treaty. The boats were partly re-engined before World War II. Their war service was limited to training, and plans to convert them to transport submarines were abandoned although they might have proved very useful in this role. On 17 August 1942 *Bass* lost half her crew when a serious fire broke out in the engine room while she was at sea. She was scuttled on 14 July 1945.

Country:	USA
Launch date:	27 December 1924
Crew:	85
Displacement:	Surfaced: 2032 tonnes (2000 tons) Submerged: 2662 tonnes (2620 tons)
Dimensions:	99.4m x 8.3m x 4.5m (326ft x 27ft 3in x 14ft 9in)
Armament:	Six 533mm (21in) torpedo tubes; one 76mm (3in) gun
Powerplant:	Two-shaft diesel engines, electric motors
Surface range:	11,118km (6000nm) at 11 knots
Performance:	Surfaced: 18 knots Submerged: 11 knots

Beta

In 1912 two small experimental submarines were built in the Venice Naval Yard for harbour surveillance and defence. They did not serve in the Italian Navy, but were given the temporary names of *Alfa* and *Beta*. Next came the 31.4-tonne (31-ton) A class of 1915-16, closely followed by the B class of which this Beta was one. She was better known as the *B1*. Only three of the class became operative, as harbour defence vessels, with three more being broken up in 1920 before completion. Harbour defence on Italy's Adriatic coast posed few problems, as virtually the only harbour of any consequence in World War I was Trieste; the Austrians, on the other hand, made use of many harbours and anchorages, and their defences were badly overstretched. Few of the naval actions in the Adriatic had a conclusive result.

Country:	Italy
Launch date:	July 1916
Crew:	20
Displacement:	Surfaced: 40 tonnes (40 tons) Submerged: 46 tonnes (46 tons)
Dimensions:	15m x 2.3m x 2.5m (49ft 7in x 7ft 8in x 8ft 3in)
Armament:	Two 450mm (17.7in) torpedo tubes
Powerplant:	Single screw petrol engine, electric motor
Surface range:	Not known
Performance:	Surfaced: 8 knots Submerged: Not known

Blaison

B*laison* was formerly the German Type IXB submarine *U-123*. She was operational from May 1940 until August 1944, when, unable for technical reasons to comply with orders to break out of Lorient and sail for Norway, she was scuttled. She was raised and commissioned into the French Navy as *Blaison* in 1947, serving until 1951. She was then placed in reserve and eventually scrapped in August 1958. Several other ex-German submarines also served with the French Navy post-war; they were the Type IXC *U510*, which surrendered at St Nazaire and became the *Bouan;* the Type VIIC *U471*, which was repared after being damaged by air attack at Toulon and commissioned as *Mille;* and the Type VIIC *U766*, which surrendered at La Pallice and became the *Laubie*. The most sought-after prize, though, was the new Type XXI.

Country:	France
Launch:	1940
Crew:	48
Displacement:	Surfaced: 1050 tonnes (1034 tons) Submerged: 1178 tonnes (1159 tons)
Dimensions:	76.5m x 6.8m x 4.7m (251ft x 22ft 4in x 15ft 5in)
Armament:	Six 533mm (21in) torpedo tubes; one 105mm (4.1in) gun
Powerplant:	Two twin-shaft diesel engines, two electric motors
Surface range:	4632km (2500nm) at 16 knots
Performance:	Surfaced: 18 knots Submerged: 7.3 knots

Brin

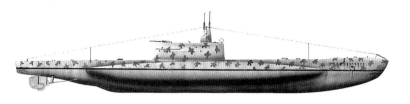

B<i>rin</i> (named after the celebrated Italian naval engineer Benedetto Brin) was one of a class of long-range submarines with a partial double hull developed from the Archimede class. A distinguishing feature of the Brin class was their tall conning tower. She was active from the beginning of Italy's involvement in World War II, initially forming part of a submarine squadron covering the approaches to the Aegean Sea. In 1941, as part of an Italian submarine group based on French Atlantic ports, she operated against Allied convoys in the sea area west of Gibraltar. Following the Italian armistice in September 1943, *Brin*, under Allied command, transferred to Ceylon and was used to train Allied anti-submarine warfare forces in the Indian Ocean, a role in which she became quite famous. She was discarded in 1948.

Country:	Italy
Launch date:	3 April 1938
Crew:	58
Displacement:	Surfaced: 1032 tonnes (1016 tons) Submerged: 1286 tonnes (1266 tons)
Dimensions:	70m x 7m x 4.2m (231ft 4in x 22ft 6in x 13ft 6in)
Armament:	Eight 533mm (21in) torpedo tubes, one 100mm (3.9in) gun
Powerplant:	Twin screw diesel engines, two electric motors
Surface range:	18,530km (10,000nm) at 10 knots
Performance:	Surfaced: 17 knots Submerged: 8 knots

Bronzo

One of 16 Acciaio-class submarines, *Bronzo* first went into action in June 1942 against the Malta-bound convoys, without success, and was one of several boats that tried to intercept the fast minelayer HMS *Welshman,* running vital supplies to the besieged island. On 12 August 1942, commanded by Lt Cdr Buldrini, she sank the freighter *Empire Hope,* already damaged and abandoned, but reported no further successes in her operational career. On 12 July 1943 she surfaced off Syracuse harbour, her captain unaware that the port was in Allied hands, and was at once raked by gunfire from the British fleet minesweepers *Boston, Cromarty, Poole* and *Seaham.* She was captured and became the British *P174*; in 1944 she was transferred to the Free French Navy and was named *Narwal.* She was scrapped in 1948.

Country:	Italy
Launch date:	28 September 1941
Crew:	46-50
Displacement:	Surfaced: 726 tonnes (715 tons) Submerged: 884 tonnes (870 tons)
Dimensions:	60m x 6.5m x 4.5m (197ft x 21ft 4in x 14ft 9in)
Armament:	Six 533mm (21in) torpedo tubes; one 100mm (3.9in) gun
Powerplant:	Two diesels, two electric motors
Surface range:	7042km (3800nm) at 10 knots
Performance:	Surfaced: 15 knots Submerged: 7.7 knots

C1 class

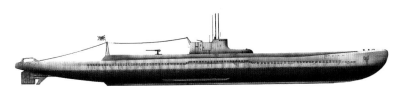

The C1 class of submarine, designated I16 in Japanese Navy service, was the product of a massive naval building programme initiated by the Japanese government after the expiry of the London Naval Treaty. There were five boats in the class (*I16, I18, I20, I22* and *I24*) and they had an extremely long radius of action, being able to remain at sea for 90 days without replenishment. At the beginning of 1943 *I16* had her 140mm (5.5in) gun removed and the number of torpedoes reduced. With special fittings she could carry a 14m (46ft) landing craft, or equipment and stores for Japanese troops on islands isolated by the Allied advance. The *I16* was sunk off the Solomon Islands by a 'hedgehog' ASW mortar salvo from a US destroyer escort group on 19 May 1944. The USS *England* of this group sank six Japanese submarines in 12 days.

Country:	Japan
Launch date:	28 July 1938
Crew:	100
Displacement:	Surfaced: 2605 tonnes (2564 tons)
	Submerged: 3761 tonnes (3701 tons)
Dimensions:	108.6m x 9m x 5m (256ft 3in x 29ft 5in x 16ft 4in)
Armament:	Eight 533mm (21in) torpedo tubes; one 140mm (5.5in) gun
Powerplant:	Two-shaft diesel/electric motors
Range:	25,928km (14,000nm)
Performance:	Surfaced: 23.5 knots
	Submerged: 8 knots

C3

All the C-class boats gave good service to the Royal Navy, and were well liked by their crews. In 1910, by which time 37 were in service, three of them, escorted by the sloop *Rosario*, were towed to the Far East to join the China Squadron at Hong Kong, a truly epic voyage for submarines in those early pioneer days, and three more went to Gibraltar. During World War I four C-class boats were sent to Russia, but were scuttled to prevent them falling into German hands in the Baltic. *C3* herself had a dramatic exit; on 23 April 1918 she was filled with high explosive and, commanded by Lt Richard D. Sandford, crept into Zeebrugge harbour and was exploded under a steel viaduct as part of the British blocking operation there. The two officers and four men aboard were picked up, although wounded; Sandford was awarded the Victoria Cross.

Country:	Britain
Launch date:	1906
Crew:	16
Displacement:	Surfaced: 295 tonnes (290 tons)
	Submerged: 325 tonnes (320 tons)
Dimensions:	43m x 4m x 3.5m (141ft x 13ft 1in x 11ft 4in)
Armament:	Two 457mm (18in) torpedo tubes
Powerplant:	Single screw petrol engine, one electric motor
Surface range:	2414km (1431nm) at 8 knots
Performance:	Surfaced: 12 knots
	Submerged: 7.5 knots

C25

The C class of submarine was the first to be produced in substantial numbers for the Royal Navy. *C25* was part of the second batch of boats (*C19* to *C38*), completed in 1909-10. Despite their limitations the C boats were active in World War I. Because of their small size, four were shipped to north Russia, broken into sections and transported overland to be reassembled for use in the Gulf of Finland. C boats were sometimes towed submerged by decoy trawlers to counter small U-boats that were harassing Britain's North Sea fishing fleet, a ruse that resulted in the destruction of two of the enemy before the Germans realised what was happening. Four C-class boats were lost in the war, and the four boats in the Gulf of Finland were eventually blown up to prevent their seizure by Communist forces after the White Russian collapse.

Country:	Britain
Launch date:	1909
Crew:	16
Displacement:	Surfaced: 295 tonnes (290 tons) Submerged: 325 tonnes (320 tons)
Dimensions:	43m x 4m x 3.5m (141ft x 13ft 1in x 11ft 4in)
Armament:	Two 457mm (18in) torpedo tubes
Powerplant:	Single screw petrol engine, one electric motor
Surface range:	2414km (1431nm) at 8 knots
Performance:	Surfaced: 12 knots Submerged: 7.5 knots

Cagni

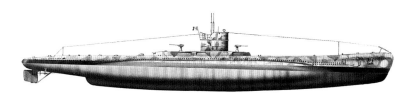

The four boats of the Ammiraglio Cagni class were the largest ever built for the Italian Navy. They were designed for commerce raiding, hence the lesser calibre of torpedo (450mm/17.7in against the more usual 533mm/21in) they carried, these being considered adequate to destroy unarmoured merchant ships. The commerce-raiding role also accounted for the unusually large number of torpedo tubes, the idea being to fire a salvo as rapidly as possible into a convoy. Each boat carried 36 torpedoes, three times as many as normal. Cagni was used as a transport submarine from 1943. Of the four Cagni-class boats, only *Cagni* herself survived World War II, being discarded in February 1948. Two were sunk by British submarines, and a third was scuttled after being damaged by a British destroyer.

Country:	Italy
Launch date:	20 July 1940
Crew:	85
Displacement:	Surfaced: 1528 tonnes (1504 tons) Submerged: 1707 tonnes (1680 tons)
Dimensions:	87.9m x 7.76m x 5.72m (200ft 5in x 17ft 7in x 13ft)
Armament:	Fourteen 450mm (17.7in) torpedo tubes; two 100mm (3.9in) guns
Powerplant:	Two diesel engines, two electric motors
Surface range:	22,236km (12,000nm) at 11 knots
Performance:	Surfaced: 17 knots Submerged: 9 knots

Casabianca

Casabianca was one of the last batch of six boats (there were 29 in all, laid down in six batches between 1925 and 1931) of the Redoutable class. They were designated First Class Submarines, but they were dogged by early misfortune when *Promethée* was lost during trials on 8 July 1932 and *Phénix* was lost in unknown circumstances in Indo-Chinese waters on 15 June 1939. Of the remainder, 11 were scuttled either at Toulon or Brest when the Germans occupied the Vichy French zone in November 1942 and others were lost during Allied attacks on Vichy French naval assets in North Africa. *Casabianca* played a proud part in the liberation of Corsica, sinking two German anti-submarine patrol boats in December 1943 and severely damaging an Italian cargo ship. Free French crews handled their boats with great skill and courage.

Country:	France
Launch date:	2 February 1935
Crew:	61
Displacement:	Surfaced: 1595 tonnes (1570 tons) Submerged: 2117 tonnes (2084 tons)
Dimensions:	92.3m x 8.2m x 4.7m (210ft x 18ft 7in x 10ft 9in)
Armament:	Nine 550mm (21.7in) and two 400mm (15.7in)torpedo tubes; one 100mm (3.9in) gun
Powerplant:	Two twin-shaft diesel engines, two electric motors
Surface range:	18,530km (10,000nm) at 10 knots
Performance:	Surfaced: 17-20 knots Submerged: 10 knots

Casma

In the mid-1960s the West German firm IKL designed a new class of submarine for the export market; this became the Type 209 class in 1967. The Peruvian Navy has six Type 209s, named *Casma, Antofagasta, Pisagua, Chipana, Islay* and *Arica*. The first pair were ordered in 1969, with two more following in August 1976 and two further boats in March 1977. The Type 209 is a single hull design with two ballast tanks, plus forward and after trim tanks. The boats are fitted with Snorkel gear and with remote machinery control. The type has an endurance of 50 days; four are in service at any one time, with two in refit or reserve. The Type 209 has a diving depth of 250m (820ft). The Type 209 was just one of a range of coastal coastal boats that were offered for export by West Germany.

Country:	Peru
Launch date:	31 August 1979
Crew:	31-35
Displacement:	Surfaced: 1122 tonnes (1105 tons) Submerged: 1249 tonnes (1230 tons)
Dimensions:	56m x 6.2m x 5.5m (183ft 9in x 20ft 3in x 18ft)
Armament:	Eight 533mm (21in) torpedo tubes
Powerplant:	Four diesels, one electric motor
Range:	4447km (2400nm) at 8 knots
Performance:	Surfaced: 10 knots Submerged: 22 knots

CB12

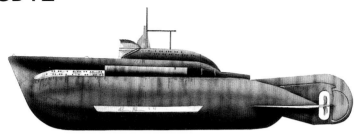

The CB programme of miniature submarines was begun in 1941 and was intended to comprise 72 vessels, but only 22 were ever laid down. They could be transported by railway and were designed for local defence. All 22 units that entered service were built by Caproni Taliedo of Milan and designed by Major Engineer Spinelli. Maximum diving depth was 55m (180ft 5in). After September 1943, *CB1* to *CB6* were transferred to Romania and subsequently scuttled (except *CB5*, which was torpedoed in Yalta harbour by a Russian aircraft). *CB8* to *CB12* were scrapped at Taranto in 1948, and the remainder were captured by the Germans while still under completion and turned over to their puppet Fascist Government in northern Italy, several being destroyed in air raids. Boats that were destroyed in this way included *CB13*, *CB14*, *CB15*, and *CB17*.

Country:	Italy
Launch date:	August 1943
Crew:	4
Displacement:	Surfaced: 25 tonnes (24.9 tons) Submerged: 36 tonnes (35.9 tons)
Dimensions:	15m x 3m x 2m (49ft 3in x 9ft 10in x 6ft 9in)
Armament:	Two 450mm (17.7in) torpedoes in external canisters
Powerplant:	Single screw diesel, one electric motor
Range:	2660km (1434nm) at 5 knots
Performance:	Surfaced: 7.5 knots Submerged: 6.6 knots

Charlie I class

The Charlie I class were the first Soviet nuclear-powered guided-missile submarines capable of launching surface-to-surface cruise missiles without having to surface first. They are similar in some respects to the Victor class, although there are visible differences that include a bulge at the bow, the almost vertical drop of the forward end of the fin, and a slightly lower after casing. The Charlie Is were all built at Gorky between 1967 and 1972; all were decommissioned by the early 1990s. *K-43* was leased to India as INS *Chakra* in January 1988 (then scrapped in 1992), and another sank off Petropavlovsk in June 1983; this vessel was later salvaged. The Charlie I carried the SS-N-15 nuclear-tipped anti-submarine missile, which has a range of 37km (20nm) and also the SS-N-7 submerged-launch anti-ship missile for pop-up surprise attacks.

Country:	Russia
Launch date:	1967
Crew:	100
Displacement:	Surfaced: 4064 tonnes (4000 tons) Submerged: 4877 tonnes (4800 tons)
Dimensions:	94m x 10m x 7.6m (308ft x 32ft 9in x 25ft)
Armament:	Eight SS-N-7 cruise missiles, six 533mm (21in) torpedo tubes
Powerplant:	Nuclear, one pressurized water reactor, one steam turbine
Range:	Unlimited
Performance:	Surfaced: 20 knots Submerged: 27 knots

Charlie II class

The Charlie II class, built between 1972 and 1980 at Gorki, was an improved Charlie I with a 9m (29ft 6in) insertion in the hull forward of the fin to house the electronics and launch systems necessary for targeting and firing the SS-N-15 and SS-N-16 weapons. In both Charlie classes, once the missiles were expended the submarine had to return to base to be reloaded. The six Charlie II boats were also armed with the SSN-9 Siren anti-ship missile, which cruises at 0.9 Mach and has a range of 110km (60nm) and can be fitted with either a nuclear (250kT) or conventional warhead. The Charlie II-class vessels were all based with the Northern Fleet, with occasional deployments to the Mediterranean. All were scrapped in the 1990s after the break up of the Soviet Union. The last in service, *K-452*, was decommissioned in 1998.

Country:	Russia
Launch date:	1973
Crew:	110
Displacement:	Surfaced: 4572 tonnes (4500 tons)
	Submerged: 5588 tonnes (5500 tons)
Dimensions:	102.9m x 10m x 7.8m (337ft 7in x 32ft 10in x 25ft 7in)
Armament:	Six 533mm (21in) and two 650mm (25.6in) torpedo tubes;
	eight cruise missiles
Powerplant:	Nuclear, one pressurized water reactor
Range:	Unlimited
Performance:	Surfaced: 20 knots
	Submerged: 26 knots

Collins

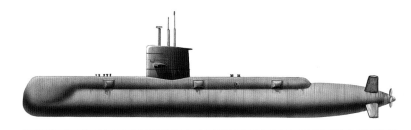

The contract for the licence production of six Swedish-designed Kockums Type 471 SSKs by the Australian Submarine Corporation, Adelaide, was signed on 3 June 1987. Fabrication work began in June 1989, the bow and midships sections of the first submarines being built in Sweden. Diving depth of the boats is 300m (984ft), and anechoic tiles were fitted to all but *Collins,* which was retrofitted. The submarines are named *Collins, Farncomb, Waller, Dechaineux, Sheean* and *Rankin*. The boats can carry 44 mines in lieu of torpedoes if required. The Collins-class submarines are very quiet, and their long range makes them very suited to operations in the southern Pacific. All are based at Fleet Base West on Garden Island off Australia's west coast, with east coast deployments. They are to be replaced by the new AUKUS class SSN, likely in the 2040s.

Country:	Australia
Launch date:	28 August 1993
Crew:	42
Displacement:	Surfaced: 3100 tonnes (3051 tons)
	Submerged: 3407 tonnes (3353 tons)
Dimensions:	77.8m x 7.8m x 7m (255ft 2in x 25ft 7in x 23ft)
Armament:	Six 533mm (21in) torpedo tubes; Sub Harpoon SSM
Powerplant:	Single shaft, diesel/electric motors
Surface range:	18,496km (9982nm) at 10 knots
Performance:	Surfaced: 10 knots
	Submerged: 20 knots

Conqueror

One of three Churchill-class nuclear-powered attack submarines (SSNs), HMS *Conqueror* was the boat that sank the Argentinian cruiser *General Belgrano* on 2 May 1982, at the start of the Falklands War. The Churchills were modified Valiant-class SSNs and were somewhat quieter in service, having benefited from the experience gained in operating the earlier boats. When the Churchills were first built their main armament was the Mk 8 anti-ship torpedo of World War II vintage, and it was a salvo of these that sank the *Belgrano*. The armament was later updated to include the Mk24 Tigerfish wire-guided dual-role (anti-ship and anti-submarine) torpedo, the Sub-Harpoon SSM and a new generation of 'smart' mines. The Churchills and their predecessors, *Valiant* and *Warspite,* were paid off in the late 1980s, following the full deployment of the Trafalgar class SSNs.

Country:	Britain
Launch date:	28 August 1969
Crew:	116
Displacement:	Surfaced: 4470 tonnes (4400 tons) Submerged: 4979 tonnes (4900 tons)
Dimensions:	86.9m x 10.1m x 8.2m (285ft x 33ft 3in x 27ft)
Armament:	Six 533mm (21in) torpedo tubes
Powerplant:	Nuclear, one pressurized water reactor
Range:	Unlimited
Performance:	Surfaced: 20 knots Submerged: 29 knots

Corallo

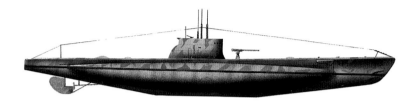

Corallo was one of ten submarines of the Perla class, all completed in 1936. Two of the class, *Iride* and *Onice*, served under the Nationalist flag in the Spanish Civil War, bearing the temporary names *Gonzalez Lopez* and *Aguilar Tablada;* during World War II *Iride*, together with a sister boat, *Ambra*, were modified to carry human torpedoes. In September 1940 *Corallo* took part in a failed attack on the aircraft carrier HMS *Illustrious* and the battleship HMS *Valiant*. During the next two years she enjoyed some small success, sinking two sailing vessels off the North African coast, but on 13 December 1942 she was sunk off Bougie by the British sloop *Enchantress*, the submarine *Porfido* being sunk in the same action by the British submarine *Tigris*. The British cruiser *Argonaut* was damaged by a torpedo.

Country:	Italy
Launch date:	2 August 1936
Crew:	45
Displacement:	Surfaced: 707 tonnes (696 tons) Submerged: 865 tonnes (852 tons)
Dimensions:	60m x 6.5m x 5m (196ft 9in x 21ft 2in x 15ft 3in)
Armament:	Six 533mm (21in) torpedo tubes, one 100mm (3.9in) gun
Powerplant:	Two diesel engines, two electric motors
Surface range:	6670km (3595nm) at 10 knots
Performance:	Surfaced: 14 knots Submerged: 8 knots

D1

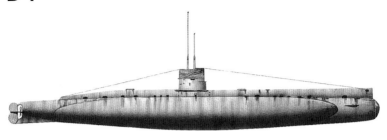

With the D-class submarines, the British made their first attempt to produce vessels that could be used on extended patrols away from coastal areas. The D class had increased displacement, diesel engines and greater internal space. Unlike earlier classes, the D class could send wireless messages as well as receive them. On the outbreak of World War I in August 1914, the eight D-class boats of the Dover-based 8th Flotilla were assigned to screen the convoys that were ferrying troops of the British Expeditionary Force across the Channel to France and to push out offensive patrols into the Heligoland Bight. The 8th Flotilla at Harwich was commanded by Commodore Roger Keyes, and as well as the D-class boats comprised nine E-class submarines. *D1* was sunk as a floating target in 1918.

Country:	Britain
Launch date:	August 1908
Crew:	25
Displacement:	Surfaced: 490 tonnes (483 tons) Submerged: 604 tonnes (595 tons)
Dimensions:	50m x 6m x 3m (163ft x 20ft 6in x 10ft 5in)
Armament:	Three 457mm (18in) torpedo tubes, one 12-pounder gun
Powerplant:	Twin screw diesel engines, electric motors
Surface range:	2038km (1100nm) at 10 knots
Performance:	Surfaced: 14 knots Submerged: 9 knots

Dagabur

Dagabur was one of the Adua class of 17 boats built for the Italian Navy in the years immediately before the outbreak of World War II. Two of the class, *Gondar* and *Scire,* were modified in 1940-41 for the transport of human torpedoes in three cylinders that were attached to the outside central part of the hull, fore and aft of the conning tower, and in December 1941 it was the *Scire* that infiltrated human torpedo crews into Alexandria harbour to make successful attacks on the British battleships *Valiant* and *Queen Elizabeth.* On 12 August 1942, *Dagabur* was part of a strong force despatched to attack the vital 'Pedestal' supply convoy to Malta; she was attempting to set up an attack on the aircraft carrier HMS *Furious* when she was rammed and sunk by the destroyer *Wolverine.*

Country:	Italy
Launch date:	22 November 1936
Crew:	45
Displacement:	Surfaced: 690 tonnes (680 tons) Submerged: 861 tonnes (848 tons)
Dimensions:	60m x 6.5m x 4m (197ft 6in x 21ft x 13ft)
Armament:	Six 533mm (21in) torpedo tubes, one 100mm (3.9in) gun
Powerplant:	Twin screw diesel engines, electric motors
Surface range:	4076km (2200nm) at 10 knots
Performance:	Surfaced: 14 knots Submerged: 8 knots

Dandolo

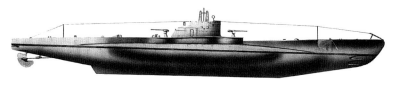

Dandolo, a long-range, single hull boat with internal ballast tanks, was one of the nine vessels of the Marcello class, among the best Italian ocean-going submarines to see service in World War II. In the late summer of 1940, *Dandolo* moved to Bordeaux with other Italian submarines to begin offensive operations in the Central Atlantic, during which she sank one ship of 5270 tonnes (5187 tons), and damaged another of 3828 tonnes (3768 tons). She remained at Bordeaux for several months during the winter of 1940-41 – during which time there were actually more Italian submarines than German U-boats operating in the Atlantic area – before returning to the Mediterranean, where she scored further successes, including the torpedoing of the cruiser *Cleopatra* in July 1943. The only boat of her class to survive the war, she was scrapped in 1947.

Country:	Italy
Launch date:	20 November 1937
Crew:	57
Displacement:	Surfaced: 1080 tonnes (1063 tons) Submerged: 1338 tonnes (1317 tons)
Dimensions:	73m x 7.2m x 5m (239ft 6in x 23ft 8in x 16ft 5in)
Armament:	Eight 533mm (21in) torpedo tubes; two 100mm (3.9in) guns
Powerplant:	Twin screw diesel engines, electric motors
Surface range:	4750km (2560nm) at 17 knots
Performance:	Surfaced: 17.4 knots Submerged: 8 knots

Daniel Boone

Although actually two classes, the 12 Benjamin Franklin-class and the 19 Lafayette-class of nuclear-powered ballistic-missile submarines (SSBN) were very similar in appearance, the main difference being that the former were built with quieter machinery outfits. The *Daniel Boone* (SSBN629) was one of the Lafayette class. As built, the first eight Lafayettes carried the 16 Polaris A2 submarine-launched ballistic missiles (SLBMs), each with a single 800kT yield warhead, but the rest were armed with the Polaris A3, which was fitted with three independently targeted warheads; this was in turn replaced by the Poseidon C3. Between September 1978 and December 1982 12 units were converted to carry the Trident I C4 SLBM. The boats were progressively deactivated as the Trident-armed Ohio-class SSBNs entered service.

Country:	USA
Launch date:	22 June 1963
Crew:	140
Displacement:	Surfaced: 7366 tonnes (7250 tons) Submerged: 8382 tonnes (8250 tons)
Dimensions:	130m x 10m x 10m (425ft x 33ft x 33ft)
Armament:	Sixteen Polaris missiles, four 533mm (21in) torpedo tubes
Powerplant:	One water-cooled nuclear reactor, turbines
Range:	Unlimited
Performance:	Surfaced: 20 knots Submerged: 35 knots

Daphné

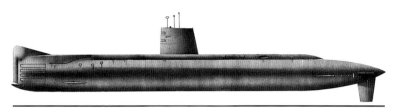

The Daphné class was designed in 1952 as a second-class ocean-going submarine to complement the larger Narval class. The boats were purposely designed with reduced speed in order to achieve a greater diving depth and heavier armament than was possible with the contemporary *Aréthuse* design of conventionally-powered hunter-killer submarines. To reduce the crew's workload the main armament was contained in 12 externally-mounted torpedo tubes, eight forward and two aft, which eliminated the need for a torpedo room and reloads. A total of 11 units was built for the French Navy. Two of the class, *Minérve* and *Eurydicé*, were lost with all hands in the western Mediterranean in 1968 and 1970 respectively. All the Daphnés had been laid up by 1990. The *Flore* went to Saudi Arabia for use as a training craft.

Country:	France
Launch date:	20 June 1959
Crew:	45
Displacement:	Surfaced: 884 tonnes (870 tons) Submerged: 1062 tonnes (1045 tons)
Dimensions:	58m x 7m x 4.6m (189ft 8in x 22ft 4in x 15ft)
Armament:	Twelve 552mm (21.7in) torpedo tubes
Powerplant:	Two diesels, two electric motors
Surface range:	8334km (4500nm) at 5 knots
Performance:	Surfaced: 13.5 knots Submerged: 16 knots

Deep Quest

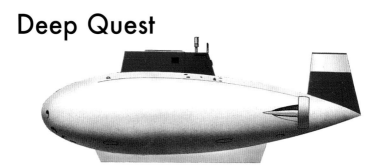

Deep Quest was the first submersible built with a fairing around a double sphere, one for the crew, the other for the propulsion unit. She works as a deep search and recovery submarine, and can descend down to 2438 metres (8000ft). Even the most advanced military submarines go down no further than 900 metres. The pressure is crushing: 60 atmospheres at 600 metres, reaching 500 on the mid-ocean bed. Pressure can exceed 1000 atmospheres in the deepest trenches: a force of almost seven tons per square inch. Vessels such as Deep Quest are vital when surveying the sea bed for cable-laying and pipeline operations. A famous vessel of similar age and type is the Alvin. It was brought to public attention in 1966 when it located and helped retrieve an H-bomb lost in the Meditereanean during World War II, when a B-52 bomber crashed.

Country:	USA
Launch date:	June 1967
Crew:	1-3
Displacement:	Surfaced: 5 tonnes (5 tons) Submerged: not known
Dimensions:	12m long (39ft 4in)
Armament:	None
Powerplant:	Twin reversible thrust motors
Operational depth:	2438m (8000ft)
Performance:	Surfaced: 4.5 knots Submerged: not known

Deepstar 4000

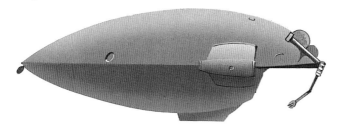

The *Deepstar 4000* was built between 1962 and 1964 by the Westinghouse Electric Corporation and the Jacques Cousteau group OFRS. The hull consists of a steel sphere with 11 openings, and she carries a wide range of scientific equipment. Pioneer vessels such as *Deepstar 4000* have lead to the discovery of hydrothermal vents, and entirely new species and ecosystems on the ocean floor. Technology has always limited the depth to which a manned submersible can operate, and during the 1980s and 1990s, great progress was made in the devel-opment of automatic or robotic deep-sea vehicles. At the turn of the twenty-first century, however, manned submersibles were increasing in their scope. The United States has developed *Deep Flight*, which is a torpedo-shaped probe designed to take one man to the bottom of the Pacific Ocean.

Country:	France
Launch date:	1965
Crew:	1
Displacement:	not known
Dimensions:	5.4m x 3.5m x 2m (17ft 9in x 11ft 6in x 6ft 6in)
Armament:	None
Powerplant:	Two fixed, reversible 5hp AC motors
Operational depth	1000m (3281ft)
Performance:	Surfaced: 3 knots Submerged: not known

Delfino

Constructed at La Spezia Naval Dockyard, *Delfino* was the first submarine built for the Italian Navy. She was rebuilt in 1902 with increased dimensions and displacement. A petrol engine was added and the conning tower enlarged. She was discarded in 1918. She was powered originally by only an electric motor, but in 1902 or thereabouts she was fitted with a petrol engine for use on the surface. Sometimes known as the *Delfino-Pullino*, she may have been laid down in the Autumn of 1889 and launched either in 1890 or 1892. The fact is that a good deal of ambiguity is attached to the various key dates in her career. What is fairly certain is that she was completed in 1892 and that she completed her first sea trials in April that year. According to official sources she was commissioned in 1896, but other sources claim that she was commissioned in April 1892.

Country:	Italy
Launch date:	1890 or 1892
Crew:	8-11
Displacement:	Surfaced: 96 tonnes (95 tons) Submerged: 108 tonnes (107 tons)
Dimensions:	24m x 3m x 2.5m (78ft 9in x 9ft 5in x 8ft 4in)
Armament:	Two 355mm (14in) torpedo tubes
Powerplant:	One petrol engine, one electric motor
Surface range:	Not known
Performance:	Surfaced: Not known Submerged: Not known

Delfino

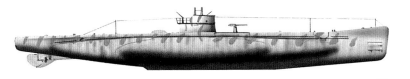

Completed in 1931, *Delfino* was one of the Squalo class of vessels, only one of which survived World War II. The Squalo-class boats were the project of General Engineer Curio Bernardis. During the first months of Italy's war she operated in the Aegean and off Crete. On 30 July 1941 she was attacked by a British Sunderland flying boat off Mersa Matruh, but succeeded in shooting it down and took four survivors prisoner. Apart from that, her operations brought no success. On 23 March 1943, she was accidentally sunk off Taranto by a collision with a pilot boat. Of her sister boats, *Narvalo* was scuttled after being damaged by British destroyers off Tripoli in January 1943, *Tricheco* was sunk by HM submarine *Upholder* off Brindisi in March 1942, and *Squalo* was discarded in 1948.

Country:	Italy
Launch date:	27 April 1930
Crew:	52
Displacement:	Surfaced: 948 tonnes (933 tons) Submerged: 1160 tonnes (1142 tons)
Dimensions:	70m x 7m x 7m (229ft x 23ft x 23ft)
Armament:	Eight 533mm (21in) torpedo tubes, one 102mm (4in) gun
Powerplant:	Two diesel engines, two electric motors
Surface range:	7412km (4000nm) at 10 knots
Performance:	Surfaced: 15 knots Submerged: 8 knots

Delta I

Until the early 1970s, the USA led the world in highly sophisticated and effective nuclear-missile submarines. Then the Russians deployed a new class of ballistic-missile submarine, the Delta I, or Murena-class SSBN which was a major improvement on the earlier Yankee class and which was armed with missiles that could outrange the American Poseidon. Each boat was armed with 12 two-stage SS-N-8 missiles. The first Delta was laid down at Severodvinsk in 1969, launched in 1971 and completed in the following year. The first of the class was paid off in 1992, three in 1993, six in 1994, one in 1995, two in 1996 and one in 1997. The remaining four boats, three of which were based with the Northern Fleet at Ostrovny and one in the Pacific at Petropavlovsk, were all decommissioned by 2004.

Country:	Russia
Launch date:	1971
Crew:	120
Displacement:	Surfaced: Submerged: 11,176 tonnes (11,000 tons)
Dimensions:	150m x 12m x 10.2m (492ft x 39ft 4in x 33ft 6in)
Armament:	Twelve missile tubes, six 457mm (18in) torpedo tubes
Powerplant:	Nuclear, two reactors
Range:	Unlimited
Performance:	Surfaced: 19 knots Submerged: 25 knots

Delta III

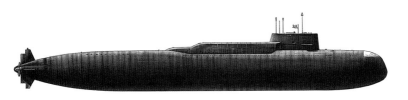

The Delta III- or Kalmar-class SSBN, completed between 1976 and 1982, had some visible differences from the earlier Delta II class from which it evolved, the most noticeable being that the missile casing was higher in order to accommodate the SS-N-18 missiles, which were longer than the SS-N-8s of the Delta II. The last of the class is the Delta IV, construction of which was first ordered in December 1975. The first of 7 boats was launched and commissioned in 1984 at Severodvinsk and the programme was completed in 1992. Larger than the Delta III, the Delta IV is a new class of submarine, which has been given the name *Delfin* (Dolfin); all still in service are with the Northern Fleet. They are being replaced with the latest Russian SSBN design, the Borei class, the first of which was commissioned in 2013.

Country:	Russia
Launch date:	1976
Crew:	130
Displacement:	Surfaced: 10,719 tonnes (10,550 tons) Submerged: 13,463 tonnes (13,250)
Dimensions:	160m x 12m x 8.7m (524.9ft x 39.4ft x 28.5ft)
Armament:	Sixteen SS-N-18 missiles; four 533mm (21in) torpedo tubes
Powerplant:	Nuclear; two pressurized water reactors, turbines
Range:	Unlimited
Performance:	Surfaced: 14 knots Submerged: 24 knots

Deutschland

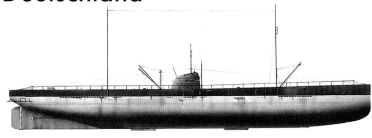

Before America's entry into the war in 1917 the Germans were quick to recognize the potential of large, cargo-carrying submarines as a means of beating the blockade imposed on Germany's ports by the Royal Navy. Two U151-class submarines, the *U151* and *U155*, were converted for mercantile use and named *Oldenburg* and *Deutschland* respectively. Both were unarmed. *Deutschland* made two commercial runs to the United States before America's involvement in the war brought an end to the venture; she was then converted back to naval use, as was *Oldenburg*. *Deutschland* was scrapped at Morecambe, England, in 1922, while *Oldenburg* was sunk as a target off Cherbourg in 1921. A third merchant conversion, *Bremen*, was lost on her first voyage in 1917, possibly mined off the Orkneys.

Country:	Germany
Launch date:	March 1916
Crew:	56
Displacement:	Surfaced: 1536 tonnes (1512 tons) Submerged: 1905 tonnes (1875 tons)
Dimensions:	65m x 8.9m x 5.3m (213ft 3in x 29ft 2in x 17ft 5in)
Armament:	None
Powerplant:	Twin screw diesel engines, electric motors
Surface range:	20,909km (11,284nm) at 10 knots
Performance:	Surfaced: 12.4 knots Submerged: 5.2 knots

Diablo

Diablo was a double-hulled ocean-going submarine developed from the previous Gato class, but was more strongly built with an improved internal layout, which increased the displacement by about 40 tonnes. She belonged to the Tench class of 50 boats, many of which were cancelled when it was realized that the Pacific war was drawing to an end. *Diablo* did not see action during World War II. In 1964, after an extensive overhaul and refit, she was transferred on loan to Pakistan from the US Navy and renamed *Ghazi*, which means Defender of the Faith. She was sunk during the 1971 war between India and Pakistan. In addition to *Ghazi*, the Pakistan Navy possessed three French Daphné-class submarines during this period; these were the *Hangor*, *Mangro* and *Shushuk*.

Country:	USA
Launch date:	30 November 1944
Crew:	85
Displacement:	Surfaced: 1890 tonnes (1860 tons) Submerged: 2467 tonnes (2420 tons)
Dimensions:	93.6m x 8.3m x 4.6m (307ft x 27ft 3in x 15ft 3in)
Armament:	Ten 533mm (21in) torpedo tubes; two 150mm (5.9in) guns
Powerplant:	Twin screw diesel engines, electric motors
Surface range:	22,518km (12,152nm) at 10 knots
Performance:	Surfaced: 20 knots Submerged: 10 knots

Diaspro

One of the ten Perla-class submarines that were all involved in the Spanish Civil War before the outbreak of World War II, *Diaspro* had a singularly undistinguished combat career, in common with many other Italian submarines. It is a fact that, with few exceptions, Italian submarine commanders failed to prosecute their operations with vigour and determination, being reluctant to attack Allied naval forces and convoys if the odds against them seemed unfavourable. What they might have achieved was revealed, doubtless to their humiliation, when the Germans stepped up the number of U-boats operating in the Mediterranean, and began to take a dramatic toll of Allied shipping from the Levant to Gibraltar, with little regard for their own danger. *Diaspro* survived the war and was stricken in 1948.

Country:	Italy
Launch date:	5 July 1936
Crew:	45
Displacement:	Surfaced: 711 tonnes (700 tons) Submerged: 873 tonnes (860 tons)
Dimensions:	60m x 6.4m x 4.6m (197ft 5in x 21ft 2in x 15ft)
Armament:	Six 533mm (21in) torpedo tubes; one 100mm (3.9in) gun
Powerplant:	Twin screw diesel engines, eletric motors
Surface range:	6670km (3595nm) at 10 knots
Performance:	Surfaced: 14 knots Submerged: 8 knots

Dolfijn

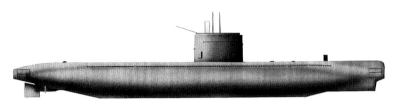

Dolfijn was one of four diesel-electric submarines completed for the Royal Netherlands Navy in the early 1960s. She was of a triple-hulled design with a maximum diving depth of nearly 304m (1000ft). Her design represented a unique solution to the problem of internal space, the hull consisting of three cylinders arranged in a triangular shape. The upper cylinder housed the crew, navigational equipment and armament, while the lower cylinders housed the machinery. Construction of the four submarines was actually authorized in 1949, but in the case of two of them was suspended for some years because of financial constraints. *Dolfijn* and *Zeehond* were laid down in December 1954, *Potvis* in September 1962 and *Tonijn* November 1962. *Dolfijn* replaced another boat of the same name, which had a distinguished career in World War II.

Country:	Netherlands
Lauch date:	20 May 1959
Crew:	64
Displacement:	Surfaced: 1518 tonnes (1494 tons) Submerged: 1855 tonnes (1826 tons)
Dimensions:	80m x 8m x 4.8m (260ft 10in x 25ft 9in x 15ft 9in)
Armament:	Eight 533mm (21in) torpedo tubes
Powerplant:	Twin screw diesels. two electric motors
Surface range:	not known
Performace:	Surfaced: 14.5 knots Submerged: 17 knots

Dolphin

The three boats of Israel's Dolphin 1 class, *Dolphin*, *Leviathan* and *Tekumah*, built by HDW in Kiel to Israeli Navy specifications, are enlarged and improved versions of the widely-exported German Type 209 class, entering service in 1999 and 2000. In 2006 Israel contracted ThyssenKrupp to build two new vessels at HDW, further upgraded with an air-independent propulsion system and classified as the Dolphin 2 class; named *Tanin* and *Rahav*, they entered service in 2014 and 2016 respectively. Both Dolphin classes carry the Popeye Turbo SLCMs, which can be nuclear-tipped, giving Israel an offshore second-strike capability. A sixth submarine, INS *Drakon*, was launched in 2023 with a much-enlarged sail that has vertical launch tubes for ballistic missiles of an unknown type. The Dolphin 1 vessels are due to be replaced by the new German-built Dakar class in the 2030s.

Country:	Germany/Israel
Launch date:	April 1996
Crew:	35
Displacement:	Surfaced: 2083 tonnes (2050 tons) Submerged: 2438 tonnes (2400 tons)
Dimensions:	97m x 8.5m x 4m (319ft 3in x 27ft 9in x 13ft 3in)
Armament:	Four 650mm (25.5in), six 533mm (21in) launch tubes; torpedoes, Popeye Turbo cruise missiles, mines
Powerplant:	Diesel-electric, single screw
Surface range:	14,816km (8000nm) at 8 knots
Performance:	Surfaced: N/A Submerged: 20 knots

Domenico Millelire

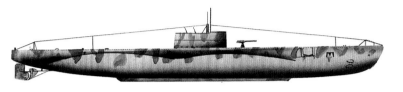

Domenico Millelire was one of four boats of the Balilla class which were all built at the Ansaldo-San Giorgio Yards. They were the first large submarines built for the Italian Navy, and they all made numerous long-range ocean cruises in the 1930s. They also participated in the Spanish Civil War on the Nationalist side, bearing false identities. The *Millelire* carried out early war patrols off Crete, and convoy protection duty in the Strait of Otranto, at the entrance to the Adriatic. She was laid up on 15 April 1941 and used as a floating oil depot, bearing the registration GR248. The class leader, *Balilla,* was also laid up in the same month and used for similar purposes. A third boat, *Antonio Sciesa,* was scuttled after being damaged in an air attack on Tobruk, while the fourth, *Enrico Toti,* was laid up in April 1943.

Country:	Italy
Launch date:	19 September 1927
Crew:	76
Displacement:	Surfaced: 1585 tonnes (1560 tons) Submerged: 2275 tonnes (2240 tons)
Dimensions:	97m x 8.5m x 4m (319ft 3in x 27ft 9in x 13ft 1in)
Armament:	Six 533mm (21in) torpedo tubes; one 102mm (4in) gun
Powerplant:	Twin screw diesels, one auxiliary motor, two electric motors
Surface range:	7401km (3800nm) at 10 knots
Performance:	Surfaced: 17.5 knots Submerged: 8.9 knots

Doris

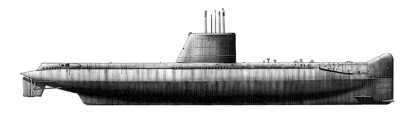

Doris was the third vessel of the Daphné class designed in 1952 to complement the larger Narval class. The design uses the double-hull construction technique with the accommodation spaces split evenly fore and aft of the sail, below which is the operations and attack centre. Crew reductions were made possible by the adoption of a modular replacement system for onboard maintenance. Of the 11 units built for France, two, *Minérve* in 1968 and *Eurydicé* in 1970, were lost with all hands while operating in the Mediterranean. The remaining boats all underwent an electronics and weapons modernization from 1970 onwards. Four units were sold to Pakistan, and in 1971 the Pakistani submarine *Hangor* sank the Indian navy frigate *Khukri* during the Indo-Pakistan War. This was the first submarine attack since World War II.

Country:	France
Launch date:	20 June 1959
Crew:	45
Displacement:	Surfaced: 884 tonnes (870 tons) Submerged: 1062 tonnes (1045 tons)
Dimensions:	58m x 7m x 4.6m (189ft 8in x 22ft 4in x 15ft)
Armament:	Twelve 552mm (21.7in) torpedo tubes
Powerplant:	Two diesels, two electric motors
Surface range:	8334km (4500nm) at 5 knots
Performance:	Surfaced: 13.5 knots Submerged: 16 knots

Dreadnought

Launched on Trafalgar Day, 21 October 1960, HMS *Dreadnought* was the Royal Navy's first nuclear-powered attack submarine (SSN), and was specifically designed to hunt and destroy hostile undersea craft. She was powered by an American S5W reactor, which was also used in the US Navy's Skipjack-class nuclear submarines; subsequent Royal Navy SSNs had British-designed nuclear plant. *Dreadnought* began sea trials in 1962. The Royal Navy carried out much pioneering work with *Dreadnought,* including proving the concept of using nuclear submarines to act as escorts for a fast carrier task group; the results of this work were made available to the US Navy, which had a close relationship with the Royal Navy at this time. Although used as a trials vessel, *Dreadnought* was a fully-capable SSN.

Country:	Britain
Launch date:	21 October 1960
Crew:	88
Displacement:	Surfaced: 3556 tonnes (3500 tons) Submerged: 4064 tonnes (4000 tons)
Dimensions:	81m x 9.8m x 8m (265ft 9in x 32ft 3in x 26ft 3in)
Armament:	Six 533mm (21in) torpedo tubes
Powerplant:	Single screw, nuclear reactor, steam turbines
Range:	Unlimited
Performance:	Surfaced: 20 knots Submerged: 30 knots

Drum

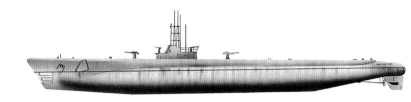

D*rum* was a double hull, ocean-going submarine with good seakeeping qualities and range. She was one of the Gato class of over 300 boats, and as such was part of the largest warship project undertaken by the US Navy. These boats, more than any others, were to wreak havoc on Japan's mercantile shipping in the Pacific war. During her first offensive patrol in April 1942 *Drum* (Lt Cdr Rice) sank the seaplane carrier *Mizuho* and two merchant ships, and later in the year she carried out vital reconnaissance work prior to the American landings in Guadalcanal. In October 1942 she sank three more ships off the east coast of Japan, and in December she torpedoed the Japanese carrier *Ryuho*. She sank a further two ships in April 1943, another in September, one in November, and three in October 1944, with another damaged. She is now a museum exhibit.

Country:	USA
Launch date:	12 May 1941
Crew:	80
Displacement:	Surfaced: 1854 tonnes (1825 tons)
	Submerged: 2448 tonnes (2410 tons)
Dimensions:	95m x 8.3m x 4.6m (311ft 9in x 27ft 3in x 15ft 3in)
Armament:	Ten 533mm (21in) torpedo tubes; one 76mm (3in) gun
Powerplant:	Twin screw diesels, electric motors
Surface range:	22236km (12,000nm) at 10 knots
Performance:	Surfaced: 20 knots
	Submerged: 10 knots

Dupuy de Lôme

Dupuy de Lôme was laid down as part of the 1913 naval construction programme. She served with the Morocco Flotilla from 1917 until the end of World War I, and was then reconstructed. Her steam engines were replaced by diesels taken from German submarines which developed 2900hp. *Dupuy de Lôme* was discarded in 1935. The submarine was named after Stanislas Charles Henri Laurent Dupuy de Lôme (1816–1885), the talented naval engineer who designed the first screw-driven warship, *Napoléon,* and the first French armoured battleship, *Gloire.* Both the French and Italians had a tendency to name their warships after engineers and statesmen, a practice not adopted by the Royal Navy. The previous warship to carry the name was a cruiser, launched in 1890 and sold to Peru in 1912.

Country:	France
Launch date:	September 1915
Crew:	54
Displacement:	Surfaced: 846 tonnes (833 tons) Submerged: 1307 tonnes (1287 tons)
Dimensions:	75m x 6.4m x 3.6m (246ft x 21ft x 11ft 10in)
Armament:	Eight 450mm (17.7in) torpedo tubes
Powerplant:	Twin screw three cylinder reciprocating steam engine; electric motors
Surface range:	10,469km (5650nm) at 10 knots
Performance:	Surfaced: 15 knots Submerged: 8.5 knots

Durbo

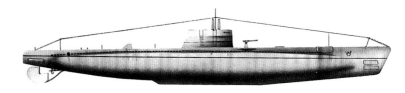

Durbo was one of the 17-strong Adua class of submarines, built in the years immediately before World War II. The Aduas might well be called the workhorses of Italy's wartime submarine fleet. Italy's entry into the conflict in June 1940 found *Durbo* at her war station in the Sicilian Channel; in July she moved to a new station off Malta, and early in September, together with two other submarines, she failed to locate and intercept a supply convoy that was approaching the island. *Durbo's* end came on 18 October 1940, when she was operating to the east of Gibraltar; located by two SARO London flying boats of No 202 Squadron RAF, she was attacked and sunk by the destroyers *Firedrake* and *Wrestler.* Another Adua-class submarine operating in the area, *Lafole,* was sunk on 20 October by the destroyers *Gallant, Griffin* and *Hotspur.*

Country:	Italy
Launch date:	6 March 1938
Crew:	45
Displacement:	Surfaced: 710 tonnes (698 tons) Submerged: 880 tonnes (866 tons)
Dimensions:	60m x 6.4m x 4m (197ft 6in x 21ft x 13ft)
Armament:	Six 533mm (21in) torpedo tubes; one 100mm (3.9in) gun
Powerplant:	Twin screw diesel engines, electric motors
Surface range:	4076km (2200nm) at 10 knots
Performance:	Surfaced: 14 knots Submerged: 7.5 knots

Dykkeren

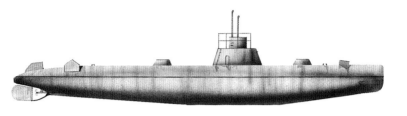

Dykkeren was built in Italy by Fiat-San Giorgio, La Spezia. She was sold to the Danish Navy in October 1909. She had many teething troubles but improved at the Copenhagen Naval Yard. In 1916 she was in collision with the Norwegian steamer *Vesta* off Bergen, and sank. Salvaged in 1917, she was broken up the following year. Although a small country, Denmark maintained a strong and efficient navy for coastal defence, and in fact possessed one of the world's first ironclads – the *Rolf Krake.* A number of ironclad warships were built for coastal defence and two were purchased which had been intended for the Confederate Navy. Denmark succeeded in maintaining its neutrality throughout the various European conflicts until 1940, when the country was occupied by German forces. She acquired four patrol submarines post-World War II.

Country:	Denmark
Launch date:	June 1909
Crew:	35
Displacement:	Surfaced: 107 tonnes (105 tons) Submerged: 134 tonnes (132 tons)
Dimensions:	34.7m x 3.3m x 2m (113ft 10in x 10ft 10in x 6ft 6in)
Armament:	Two 457mm (18in) torpedo tubes
Powerplant:	Twin screw petrol engine, one electric motor
Surface range:	185km (100nm) at 12 knots
Performance:	Surfaced: 12 knots Submerged: 7.5 knots

E11

Completed between 1913 and 1916, the E-class submarines ran to 55 hulls whose construction, once war was declared, was shared between 13 private yards. They fell into five major groups, differences being primarily in torpedo layout and the adaptation of six boats to carry 20 mines in place of their amidships tubes. *E11*, under the command of the talented Lt Cdr Martin Nasmith, was arguably the most famous of them all; operating in the Dardanelles area she scored many successes, including the sinking of the Turkish battleship *Hairredin Barbarossa*. Many RN submariners who rose to high rank learned their trade in E-class boats. For operations in the Dardanelles, the British submarines adopted a blue camouflage to conceal themselves in the shallow, clear waters. The class was also active in the North Sea and the Baltic. In all, 22 were lost.

Country:	Britain
Launch date:	1913
Crew:	30
Displacement:	Surfaced: 677 tonnes (667 tons) Submerged: 820 tonnes (807 tons)
Dimensions:	55.17m x 6.91m x 3.81m (181ft x 22ft 8in x 12ft 6in)
Armament:	Five 457mm (18in) torpedo tubes; one 12-pounder gun
Powerplant:	Two twin-shaft diesel engines, two electric motors
Surface range:	6035km (3579nm)
Performance:	Surfaced: 14 knots Submerged: 9 knots

E20

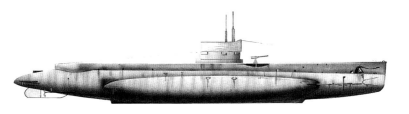

The excellent combat radius of the E-class boats enabled them to mount extended patrols in enemy waters from relatively distant bases. *E20*'s hunting ground was the Sea of Marmara, where she was sunk on 5 November 1915 by the German submarine *UB14* – the first instance of a submarine sinking another. *E20* had fallen into a trap; some time earlier, in July 1915, the Germans had sunk the small French submarine *Mariotte,* and had captured some documents which revealed the places in the Marmara Sea where British and French submarines habitually made rendezvous. Despite setbacks such as this, and despite treacherous operating conditions, the British submariners achieved considerable success in the area. At the end of the war, the boats were assembled at Malta to await disposal.

Country:	Britain
Launch date:	June 1915
Crew:	30
Displacement:	Surfaced: 677 tonnes (667 tons) Submerged: 820 tonnes (807 tons)
Dimensions:	55.6m x 4.6m x 3.8m (182ft 5in x 15ft x 12ft 6in)
Armament:	Five 457mm (18in) torpedo tubes; one 12-pounder gun
Powerplant:	Twin screw diesel engines, electric motors
Surface range:	6035km (3579nm) at 10 knots
Performance:	Surfaced: 14 knots Submerged: 9 knots

Echo

The five Echo-class SSNs were originally built at Komsomolsk in the Soviet Far East in 1960-62 as Echo I-class missile submarines (SSGNs). Armed with six tubes for the SS-N-3C Shaddock strategic cruise missile, they lacked the fire control and guidance radars of the later Echo II class, 29 of which were built. All but five of the Echo IIs served in the Northern Fleet. As the Soviet ballistic-missile submarine force was built up the need for these interim missile boats diminished, and they were converted to anti-ship attack SSNs between 1969 and 1974. The conversion involved the removal of the Shaddock tubes, the plating over and streamlining of the hull to reduce the underwater noise caused by the tube system, and modification of the sonar systems. The Echo boats were withdrawn in the 1980s.

Country:	Russia
Launch date:	1960
Crew:	90
Displacement:	Surfaced: 4572 tonnes (4500 tons) Submerged: 5588 tonnes (5500 tons)
Dimensions:	110m x 9m x 7.5m (360ft 11in x 29ft 6in x 24ft 7in)
Armament:	Six 533mm (21in) and two 406mm (16 in) torpedo tubes
Powerplant:	One pressurized water reactor, two steam turbines
Range:	Unlimited
Performance:	Surfaced: 20 knots Submerged: 28 knots

Enrico Tazzoli

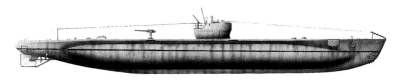

One of the four-strong Calvi class, *Enrico Tazzoli* was completed in 1936. She took part in the Spanish Civil War, and served in the Mediterranean in the early months after Italy's entry into World War II. In 1940 she transferred to the Atlantic, scoring an early success in October by sinking a 5217-tonne (5135-ton) freighter off the Portuguese coast. In December 1941 she was involved in a rescue operation in the Atlantic, ferrying part of the crew of the sunken commerce raider *Atlantis* to St Nazaire. In 1942 she was refitted to transport supplies to Japan. She left Bordeaux on May 1943 and was never heard of again, being lost somewhere in the Bay of Biscay along with another Italian submarine, *Barbarigo*. Three more Italian boats that set out on a similar mission reached Sabang and Singapore without incident.

Country:	Italy
Launch date:	14 October 1935
Crew:	77
Displacement:	Surfaced: 1574 tonnes (1500 tons) Submerged: 2092 tonnes (2060 tons)
Dimensions:	84.3m x 7.7m x 5.2m (276ft 6in x 25ft 3in x 17ft)
Armament:	Eight 533mm (21in) torpedo tubes; two 120mm (4.7in) guns
Powerplant:	Twin screw diesel engines; electric motors
Surface range:	19,311km (10,409nm) at 10 knots
Performance:	Surfaced: 17 knots Submerged: 8 knots

Enrico Tazzoli

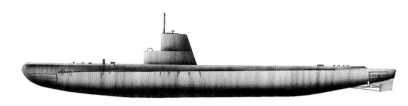

Enrico Tazzoli was formerly the US submarine *Barb*, completed in 1943 as one of the vast World War II Gato class. She transferred to the Italian Navy in 1955 after conversion to the Guppy snorkel, which included a modified structure and 'fairwater' for better underwater performance. She carried 254 tonnes (250 tons) of fuel oil, enough for 19,311km (10,409nm) at 10 knots. The Italian Navy operated a number of former US Navy oceangoing submarines at this time; the others were the *Alfredo Cappellini* (ex-USS *Capitaine); Evangelista Torricelli* (ex-USS *Lizardfish); Francesco Morosini* (ex-USS *Besugo*); and *Leonardo da Vinci,* formerly the USS *Dace.* The *Tazzoli* and *da Vinci*, both of which were on extended loan from the USA, formed the backbone of the Italian submarine fleet during this period.

Country:	Italy
Launch date:	2 April 1942
Crew:	80
Displacement:	Surfaced: 1845 tonnes (1816 tons)
	Submerged: 2463 tonnes (2425 tons)
Dimensions:	94m x 8.2m x 5m (311ft 3in x 27ft x 17ft)
Armament:	Ten 533mm (21in) torpedo tubes
Powerplant:	Twin screw diesels, electric motors
Surface range:	19,311km (10,409nm) at 10 knots
Performance:	Surfaced: 20 knots
	Submerged: 10 knots

Enrico Toti

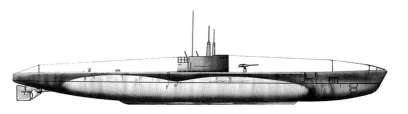

One of the Balilla class of large submarines, *Enrico Toti* was a long-range vessel with a diving depth of 90m (295ft). She served in the Spanish Civil War, and the early weeks of Italy's war found her operating against French shipping traffic evacuating troops and equipment from France to Algeria. Afterwards, she moved to a new operational area south of Crete. On 15 October 1940, off Calabria, she came upon the surfaced British submarine *Rainbow* and sank her in a gun duel. In the summer of 1942 it was decided that *Toti* was too large to operate offensively in the Mediterranean, and she spent some time in the transport role before being laid up in April 1943. Submarines of this class could carry a maximum of 16 torpedoes, and the class leader, *Balilla*, was equipped to carry four mines.

Country:	Italy
Launch date:	14 April 1928
Crew:	76
Displacement:	Surfaced: 1473 tonnes (1450 tons) Submerged: 1934 tonnes (1904 tons)
Dimensions:	87.7m x 7.8m x 4.7m (288ft x 25ft 7in x 15ft 5in)
Armament:	Six 533mm (21in) torpedo tubes; one 120mm (4.7in) gun
Powerplant:	Twin screw diesels, electric motors
Surface range:	7041km (3800nm) at 10 knots
Performance:	Surfaced: 17.5 knots Submerged: 9 knots

Enrico Toti

Enrico Toti was the lead boat in a class of four which were the first submarines to be built in Italy since World War II. The design was revised several times, and a coastal hunter-killer type intended for shallow and confined waters was finally approved. For these operations the boats' relatively small size and minimum sonar cross-section were a great advantage. The main armament carried was the Whitehead Motofides A184 wire-guided torpedo; this was a dual ASW/anti-ship weapon with an active/passive acoustic homing head that featured enhanced ECCM to counter enemy decoys. With a range in the order of 25km (13.5nm), the weapon would have proved effective in an ambush situation at natural 'choke points' against much larger opponents such as Russian SSNs or SSGNs.

Country:	Italy
Launch date:	12 March 1967
Crew:	26
Displacement:	Surfaced: 532 tonnes (524 tons)
	Submerged: 591 tonnes (582 tons)
Dimensions:	46.2m x 4.7m x 4m (151ft 7in x 15ft 5in x 13ft)
Armament:	Four 533mm (21in) torpedo tubes
Powerplant:	Single screw diesel engine, electric motors
Surface range:	5556km (3000nm) at 5 knots
Performance:	Surfaced: 14 knots
	Submerged: 15 knots

Entemedor

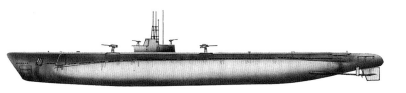

Formerly named the *Chickwick, Entemedor* was a double-hulled, ocean-going submarine of the large World War II Gato class. Fuel tanks, containing up to 480 tonnes (472 tons) were situated in the central double hull section. Maximum diving depth was 95m (312ft). *Entemedor* was transferred to Turkey in 1973. During the years of the Cold War, Turkey, a vital component of NATO's Southern Flank, had a navy that was an extraordinary collection of obsolescent American warships. In the late 1960s it included ten submarines of the Gato class, all of which had been loaned by the US under the terms of the NATO Mutual Assistance Pact. All had been modified to carry up-to-date equipment, including the 'guppy snorkel' in some cases. The *Entemedor* was named *Preveze* in Turkish service. The development of the Turkish navy has mirrored that of its main rival, Greece.

Country:	Turkey
Launch date:	17 December 1944
Crew:	80
Displacement:	Surfaced: 1854 tonnes (1825 tons) Submerged: 2458 tonnes (2420 tons)
Dimensions:	95m x 8.3m x 4.6m (311ft 9in x 27ft 3in x 15ft 3in)
Armament:	Ten 533mm (21in) torpedo tubes; one 127mm (5in) gun
Powerplant:	Twin screw diesel engines, electric motors
Surface range:	20,372km (11,000nm) at 10 knots
Performance:	Surfaced: 20 knots Submerged: 8.7 knots

Ersh (SHCH 303)

The Ersh (Pike) class of 88 boats, to which *Shch-303* belonged, were coastal submarines with a single hull and a maximum diving depth of 90m (295ft). Thirty-two were lost during World War II, but the survivors remained in service with the Soviet Navy until the mid-1950s. *Shch-303* operated in the Baltic, which was heavily mined and where most of the losses occurred; some of the Russian boats were sunk by Finnish submarines. The Russian submarines presented a great danger to the Germans, despite inflated claims of shipping sunk, and had to be guarded against at a cost that was unwelcome when German naval forces were badly needed elsewhere. *Shch-303*'s captain, I. V. Travkin, claimed to have sunk two large vessels in the Baltic, but the claim was never substantiated. *Shch-303* survived the war and was scrapped in 1958.

Country:	Russia
Launch date:	16 November 1931
Crew:	45
Displacement:	Surfaced: 595 tonnes (586 tons)
	Submerged: 713 tonnes (702 tons)
Dimensions:	58.5m x 6.2m x 4.2m (192ft x 20ft 4in x 13ft 9in)
Armament:	Six 533mm (21in) torpedo tubes, two 45mm (1.8in) guns
Powerplant:	Twin screw diesel engines, electric motors
Surface range:	11,112km (6000nm) at 8 knots
Performance:	Surfaced: 12.5 knots
	Submerged: 8.5 knots

Espadon

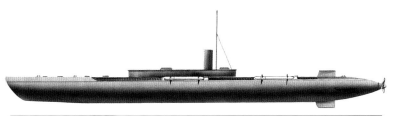

E_spadon_ (Swordfish) was one of France's earliest submarines, and was very much an experimental craft, being used for various trials. The big problem facing early submarine designers was how to propel the boat. Trials with compressed air were successful, but the storage space this required was too great to achieve reasonable speed and range. By the 1880s steam power was being used on the surface, the machinery being shut down in order to engage the newly introduced electric motors for submersion. _Espadon_ was one of this type of submarine. She was removed from the effective list in 1919. The French were very innovative in their early submarine designs, and were not afraid to experiment with novel forms of propulsion and other technological aspects of underwater craft.

Country:	France
Launch date:	September 1901
Crew:	30
Displacement:	Surfaced: 159 tonnes (157 tons)
	Submerged: 216 tonnes (213 tons)
Dimensions:	32.5m x 3.9m x 2.5m (106ft 8in x 12ft 10in x 8ft 2in)
Armament:	Four 450mm (17.7in) torpedoes
Powerplant:	Single screw triple expansion steam engine; electric motor
Surface range:	1111km (600nm) at 8 knots
Performance:	Surfaced: 9.75 knots
	Submerged: 8 knots

Espadon

The French produced some very capable submarines in the inter-war years. *Espadon* (Swordfish) belonged to the Requin (Shark) class of nine minelaying submarines that were designated First Class Submarines by the French Admiralty. They were heavily armed, with four bow, two stern and two twin torpedo tubes mounted in containers in the upper hull. All ships of the class were modernized, undergoing a complete refit of hull and machinery between 1935 and 1937. Eight of the group were lost during World War II. *Requin, Dauphin, Phoque* and *Espadon* were seized by the Italians at Bizerta on 8 December 1942; *Espadon* was towed to Castellamare di Stabia and was given the designation FR114 by her captors, but was not commissioned into the Italian Navy. She was captured by the Germans and scuttled on 13 September 1943.

Country:	France
Launch date:	28 May 1926
Crew:	54
Displacement:	Surfaced: 1168 tonnes (1150 tons) Submerged: 1464 tonnes (1441 tons)
Dimensions:	78.2m x 6.8m x 5m (256ft 9in x 22ft 5in x 16ft 9in)
Armament:	Ten 533mm (21in) torpedo tubes; one 100mm (3.9in) gun
Powerplant:	Twin screw diesel engines; electric motors
Surface range:	10,469km (5650nm) at 10 knots
Performance:	Surfaced: 15 knots Submerged: 9 knots

Ettore Fieramosca

When Italy entered the war in June 1940 no fewer than 84 submarines were operational; the remainder of the 150 boats commissioned were either in refit or undergoing trials. They were markedly inferior to all German U-boats, comfort playing too great a part in their design. Their conning towers were overlarge partly because they contained a big, well-equipped galley! In the main they were large, good-looking vessels, but they were slow to dive, clumsy when submerged and poorly equipped. *Ettore Fieramosca* was no exception. She was in a class of her own, designed for long-range ocean patrols. To this end, provision was made for her to carry a small reconnaissance aircraft in a hangar at the rear of the conning tower, but the aircraft was never embarked. The submarine was laid up in March 1941.

Country:	Italy
Launch date:	April 1929
Crew:	78
Displacement:	Surfaced: 1580 tonnes (1556 tons) Submerged: 1996 tonnes (1965 tons)
Dimensions:	84m x 8.3m x 5.3m (275ft 7in x 27ft 3in x 17ft 5in)
Armament:	Eight 533mm (21in) torpedo tubes; one 120mm (4.7in) gun
Powerplant:	Twin screw diesel engines; electric motors
Surface range:	9260km (5000nm) at 9 knots
Performance:	Surfaced: 19 knots Submerged: 10 knots

Euler

Thanks to a small number of far-sighted naval officers and politicians, France had built up a formidable submarine fleet when World War I broke out; indeed, by comparison with Germany and Great Britain, her submarine arm was extraordinarily large. Much of the reason for this lay in the fact that she felt it necessary to keep pace in naval developments with her neighbour, Great Britain. As she could not match the latter in capital ship strength her naval planners tended to concentrate on building submarines and torpedo boats – in other words, vessels that could inflict massive damage on larger ships at a relatively low cost. *Euler* formed part of a large class of 16 boats. Her submerged range was 160km (86nm) at 5 knots. She was removed from the effective list in the 1920s.

Country:	France
Launch date:	October 1912
Crew:	35
Displacement:	Surfaced: 403 tonnes (397 tons)
	Submerged: 560 tonnes (551 tons)
Dimensions:	52m x 5.4m x 3m (171ft x 17ft 9in x 10ft 3in)
Armament:	One 450mm (17.7in) torpedo tube, four drop collars,
	two external cradles
Powerplant:	Twin screw diesel engines, electric motors
Surface range:	3230km (1741nm) at 10 knots
Performance:	Surfaced: 14 knots
	Submerged: 7 knots

Eurydice

Eurydice was a double-hulled, medium-displacement submarine with an operational diving depth of 80m (262ft). She formed part of a class of 26 second-class boats built between 1925 and 1934 and was in fact one of a batch of three built at the Normand-Fenaux yard. When Italy entered the war in June 1940 *Eurydice* was at Oran, and immediately began defensive patrols off Gibraltar together with other French submarines. This was in accordance with an Anglo-French naval agreement under which the French Navy had responsibility for the defence of the western Mediterranean, an agreement that came to nothing with the Franco-German Armistice of June 1940. *Eurydice* was scuttled at Toulon on 27 November 1942 along with many other French warships, shortly before the port was occupied by II SS Panzer Corps in Operation Lila.

Country:	France
Launch date:	May 1927
Crew:	41
Displacement:	Surfaced: 636 tonnes (626 tons) Submerged: 800 tonnes (787 tons)
Dimensions:	65.9m x 4.9m x 4m (216ft 2in x 16ft x 13ft 1in)
Armament:	Seven 533mm (21in) torpedo tubes
Powerplant:	Twin screw diesel engines; electric motors
Surface range:	6485km (3500nm) at 7.7 knots
Performance:	Surfaced: 14 knots Submerged: 7.5 knots

Evangelista Torricelli

Evangelista Torricelli was formerly the US ocean-going submarine *Lizardfish* of the vast Gato class, and was originally to have been named *Luigi Torelli*. She was handed over to Italy on 5 March 1966, along with two sister vessels, the *Alfredo Cappellini* (ex-USS *Capitaine*) and *Francesco Morosini* (ex-USS *Besugo*). The United States supplied many former Gato-class submarines to its Allies in the years of the Cold War; all had their 100mm (3.9in) guns removed and were upgraded. Their long range and reliability made them very attractive vessels, as indeed did their war record in the Pacific, which was outstanding. The *Morosini* was the first to be discarded, in November 1975, followed by the *Toricelli* in 1976 and the *Cappellini* in 1977. *Toricelli* was used for experimental work in her latter years.

Country:	Italy
Launch date:	July 1944
Crew:	85
Displacement:	Surfaced: 1845 tonnes (1816 tons)
	Submerged: 2463 tonnes (2425 tons)
Dimensions:	95m x 8.2m x 5m (311ft 6in x 27ft x 17ft)
Armament:	Ten 533mm (21in) torpedo tubes
Powerplant:	Twin screw diesel engines, electric motors
Surface range:	22,518km (12,152nm) at 10 knots
Performance:	Surfaced: 20 knots
	Submerged: 10 knots

Explorer

Explorer and her sister *Excalibur* were two experimental submarines ordered from Vickers-Armstrong by the Royal Navy. The streamlined hull was designed to operate at high underwater speeds, which were made possible by using high test peroxide similar to that used in the effective German type XXI submarines built towards the end of World War II. *Explorer* was the first submarine to be launched for the Royal Navy since the completion of the A class of 1948. The vessels yielded much valuable data on future hull design connected with Britain's first-generation nuclear submarines, which were then at the initial concept phase; work on the design of the first American and Russian nuclear submarines was already well advanced in 1954, when the experimental British craft were launched.

Country:	Britain
Launch date:	March 1954
Crew:	70
Displacement:	Surfaced: 792 tonnes (780 tons) Submerged: 1016 tonnes (1000 tons)
Dimensions:	68.7m x 4.8m x 5.5m (225ft x 15ft 8in x 18ft 2in)
Armament:	None
Powerplant:	Twin screw diesel engines, hydrogen peroxide
Surface range:	Not known
Performance:	Surfaced: 20 knots Submerged: 25 knots

F1

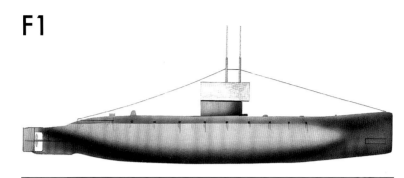

F<i>1</i> was one of a class of three vessels which were among the last coastal-defence submarines built for the Royal Navy, due to the Admiralty's decision to adopt an offensive policy by building ocean-going submarines of great range. The policy was encouraged by Winston Churchill, who became First Lord of the Admiralty in 1911 and observed that of the 57 submarines then in service with the Royal Navy, only two D-class boats were capable of operating at anything more than a short distance from Britain's shores. By the end of 1914, it had become clear that enemy submarine commanders did not intend to pursue British ships into their harbours. *F1* was laid down in 1913, and all three boats in the class saw extensive service during World War I. A proposed group of boats in the same class was cancelled in 1914. *F1* was broken up in 1920.

Country:	Britain
Launch date:	March 1915
Crew:	20
Displacement:	Surfaced: 368 tonnes (363 tons) Submerged: 533 tonnes (525 tons)
Dimensions:	46m x 4.9m x 3.2m (151ft x 16ft x 10ft 6in)
Armament:	Three 457mm (18in) torpedo tubes
Powerplant:	Twin screw diesel engines; electric motors
Surface range:	5556km (3000nm) at 9 knots
Performance:	Surfaced: 14 knots Submerged: 8.7 knots

F1

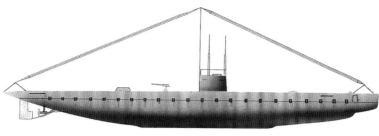

F1 and her sisters were improved versions of the Medusa class. They were able to dive faster and carried two periscopes (one for search, the other for attack) as well as a gyrocompass and the newly invented Fessenden submarine-signalling apparatus. *F1* was removed from the effective list in June 1930. Experience with craft such as *F1* should have made the Italian submarine service a sound and effective service, but between the wars it seems to have existed more for show than action. There is scant evidence of realistic pre-World War II submarine exercises, and the very special teamwork demanded of an aggressive submarine crew appears to have been lacking. In both world wars, it was the Italian torpedo-boat crews who were the most consistently aggressive in combat. On one occasion, they made a gallant attempt to attack British shipping in Malta.

Country:	Italy
Launch date:	April 1916
Crew:	54
Displacement:	Surfaced: 226 tonnes (262 tons) Submerged: 324 tonnes (319 tons)
Dimensions:	45.6m x 4.2m x 3m (149ft 7in x 13ft 9in x 10ft)
Armament:	Two 450mm (17.7in) torpedo tubes; one 76mm (3in) gun
Powerplant:	Twin screw diesel engines; electric motors
Surface range:	2963km (1600nm) at 8.5 knots
Performance:	Surfaced: 12.5 knots Submerged: 8.2 knots

F4

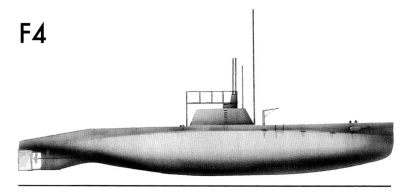

F*4* and her three sisters were similar to, and contemporaries of, the E class, where a tendency towards a smaller type of submarine had originated. All the E- and F-class boats were withdrawn from service in 1915 for re-engineering. *F4* left Honolulu harbour on 25 March 1915 for a short trial run, but she never returned. She was located at a depth of 91 metres (300ft) just off Pearl Harbor, well beyond the depth from which such a vessel had hitherto been successfully raised. Five months later, however, American salvage crews achieved the seemingly impossible and brought *F4* to the surface, setting up a new world deep-sea diving record in the process. In the years to come America was always to be at the forefront of such salvage operations, with specially-designed deep submergence craft.

Country:	USA
Launch date:	January 1912
Crew:	35
Displacement:	Surfaced: 335 tonnes (330 tons) Submerged: 406 tonnes (400 tons)
Dimensions:	43.5m x 4.7m x 3.7m (142ft 9in x 15ft 5in x 12ft 2in)
Armament:	Four torpedo tubes
Powerplant:	Twin screw diesels, electric motors
Surface range:	4260km (2300nm) at 11 knots
Performance:	Surfaced: 13.5 knots Submerged: 5 knots

Faa di Bruno

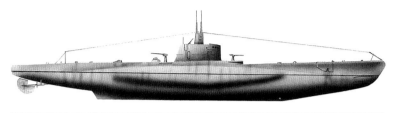

Derived from the Glauco class, the *Commandante Faa di Bruno* was a long-range, single-hulled, ocean-going boat with internal ballast tanks designed by Bernardis. There were two boats in the class, the other being the *Commandante Cappellini*. In November 1940 she was sunk in the North Atlantic through an unknown cause, but possibly by the British destroyer HMS *Havelock*. Her sister, *Cappellini,* had an interesting career. Converted to the transport role, she was captured by the Japanese at Sabang in September 1943, following the Italian armistice, and turned over to the Germans. By 1944, long-range submarines provided the only means of shipping raw materials between Germany and Japan. When Germany capitulated in May 1945 the Japanese seized her again and gave her the serial *I-503*. She ended the war at Kobe, Japan, and was scrapped in 1946.

Country:	Italy
Launch date:	18 June 1939
Crew:	58
Displacement:	Surfaced: 1076 tonnes (1060 tons) Submerged: 1334 tonnes (1313 tons)
Dimensions:	73m x 7m x 5m (239ft 6in x 23ft 7in x 16ft 9in)
Armament:	Eight 533mm (21in) torpedo tubes; two 100mm (3.9in) guns
Powerplant:	Twin screw diesel engines; electric motors
Surface range:	13,890km (7500nm) at 9.4 knots
Performance:	Surfaced: 17.4 knots Submerged: 8 knots

Farfadet

Farfadet and her three sisters relied solely on a set of accumulators for power, which gave them a range of only 218.5km (118nm) at 5.3 knots surfaced and 53km (28.5nm) at 4.3 knots submerged. The four torpedoes were carried externally in cradles aft of the conning tower. As a result of the conning tower hatch being left open, *Farfadet* sank at Bizerta on 6 July 1905, with the loss of 14 lives. She was subsequently raised and recommissioned as *Follet* in 1909. She was removed from the effective list in 1913. *Farfadet* was yet another example of French attempts to find alternative, cleaner power sources for their early submarines. French designers were extremely safety-conscious, and refused for the most part to countenance volatile petrol engines. The initial cost of each boat in this class was £32,000 (US$51,200).

Country:	France
Launch date:	May 1901
Crew:	25
Displacement:	Surfaced: 188 tonnes (185 tons) Submerged: 205 tonnes (202 tons)
Dimensions:	41.3m x 2.9m x 2.6m (135ft 6in x 9ft 6in x 8ft 6in)
Armament:	Four 450mm (17.7in) torpedo tubes
Powerplant:	Single screw, electric motors
Surface range:	218.5km (118nm) at 4.3 knots
Performance:	Surfaced: 6 knots Submerged: 4.3 knots

Fenian Ram

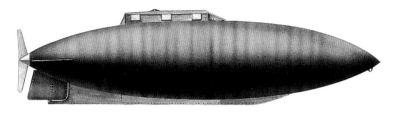

The Irish-American inventor John Philip Holland designed submarines in America, the original idea being to use them to destroy the hated British fleet on behalf of his fellow Fenians. Ironically, his business acumen overcame his nationalist sympathies and he sold his most successful design to the Royal Navy. Twenty years earlier, the *Fenian Ram* was built for the Fenian Society by the Delamater Iron Works, New York. In 1883 she was towed to Newhaven under great secrecy, so that her crew of three could familiarize themselves with her handling. The vessel was exhibited at Madison Square Gardens in 1916 in order to raise funds for the Irish uprising that took place that year. In 1927 she was housed in West Side Park, New York. John Holland has a firm place in history as the designer of the first practical submarines.

Country:	USA
Launch date:	May 1881
Crew:	3
Displacement:	Surfaced: 19 tonnes (19 tons) Submerged: Not known
Dimensions:	9.4m x 1.8m x 2.2m (31ft x 6ft x 7ft 3in)
Armament:	One 228mm (9in) gun
Powerplant:	Single screw petrol engine
Surface range:	Not known
Performance:	Surfaced: Not known Submerged: Not known

Ferraris

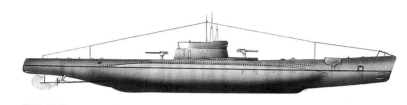

Ferraris and her sister *Galilei* were long range vessels with a partial double hull. Both boats took part in the Spanish Civil War, as did two others of the class, *Archimede* and *Torricelli*. These two were transferred to Spain in Spring 1937, being respectively named *General Sanjurjo* and *General Mola*. *Ferraris* and *Galilei* were stationed in the Red Sea at the time of Italy's entry into World War II. Galilei was captured in the Red Sea on 19 October 1940 after a surface battle with the British armed trawler *Moonstone* in which nearly all the Italian officers were killed, and the rest of the crew inside the boat were poisoned by emissions from the air-conditioning system. She was designated *X2* by the British and used for training. She was sunk in the North Adriatic by the British destroyer *Lamerton* in October 1941.

Country:	Italy
Launch date:	August 1934
Crew:	55
Displacement:	Surfaced: 1000 tonnes (985 tons) Submerged: 1279 tonnes (1259 tons)
Dimensions:	70.5m x 6.8m x 4m (231ft 4in x 22ft 4in x 13ft 1in)
Armament:	Eight 533mm (21in) torpedo tubes; two 100mm (3.9in) guns
Powerplant:	Twin screw diesel engines
Surface range:	19,446km (10,500nm) at 8 knots
Performance:	Surfaced: 17 knots Submerged: 8.5 knots

Ferro

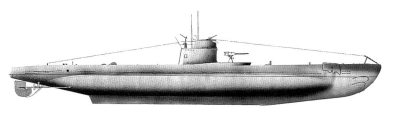

F_erro_ was designed as a medium-sized boat of the Flutto class, II series, with a partial double hull, and was a development of the Argo class with slightly increased dimensions and a reduced conning tower. She was laid down on 2 June 1943 at Cantieri Riuniti Dell'Adriatico, Monflacone, but was seized by German forces in September 1943 and designated _UIT 12_. The vessel was never launched, as she was destroyed while still on the slips in May 1945. There were to have been 25 boats in the class, but only one, the _Bario_, was ever commissioned. Damaged in an air attack, and scuttled by the Germans, she was refloated after the war and entered service under the new name of _Pietro Calvi_ on 16 December 1961. One other boat, the _Vortice_, was modified post-war, but did not enter service.

Country:	Italy
Launch date:	Not launched
Crew:	50
Displacement:	Surfaced: 1130 tonnes (1113 tons) Submerged: 1188 tonnes (1170 tons)
Dimensions:	64m x 6.9m x 4.9m (210ft 7in x 22ft 11in x 16ft 2in)
Armament:	Six 533mm (21in) torpedo tubes; one 100mm (3.9in) gun
Powerplant:	Twin screw diesel engines, electric motors
Surface range:	6670km (3600nm) at 12 knots
Performance:	Surfaced: 16 knots Submerged: 8 knots

Filippo Corridoni

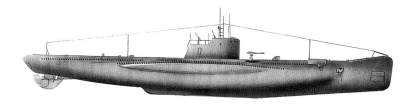

Filippo Corridoni was a short-range minelayer, one of two submarines – the other was the *Brigadin* – developed from the Pisano class. She was used mainly to transport supplies during World War II. The vessels carried two tubes for launching mines; between 16 and 24 of the latter could be carried, depending on the type. Some 17 different types of mine were used by the Italian Navy in World War II, including one known as the Coloniale P125 developed specifically for use in warm seas. The Italians also used German mines from 1941; these were generally more effective than the Italian models. The Italian Navy placed great faith in mines, and many boats of their submarine fleet were converted to the minelaying role. The *Filippo Corridoni* was removed from the effective list in 1948, as was the *Brigadin*.

Country:	Italy
Launch date:	March 1930
Crew:	55
Displacement:	Surfaced: 996 tonnes (981 tons) Submerged: 1185 tonnes (1167 tons)
Dimensions:	71.5m x 6m x 4.8m (234ft 7in x 20ft 2in x 15ft 9in)
Armament:	Four 533mm (21in) torpedo tubes; one 102mm (4in) gun; up to 24 mines
Powerplant:	Twin screw diesel engines, electric motors
Surface range:	16,668km (9000nm) at 8 knots
Performance:	Surfaced: 11.5 knots Submerged: 7 knots

Fisalia

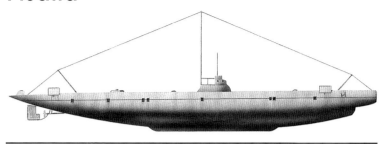

$\mathbf{F}$*isalia* was one of a class of eight boats that were the first Italian submarines to have diesel engines. They were good sea boats with excellent manoeuvrability. *Fisalia* was laid down in October 1910 and completed in September 1912. She served in the Adriatic during World War I, and was discarded in 1918. One or two Italian boats deployed to the Dardanelles in 1915, but it was the British and French submarines that excelled in this area, particularly the Royal Navy's E class. The Italian boats of the period were generally confined to the Adriatic because of their short range; they were also lacking in offensive armament, and so were used mainly for harbour and coastal defence, leaving offensive action to the torpedo boats. As a consequence, the latter scored most of the successes.

Country:	Italy
Launch date:	February 1912
Crew:	40
Displacement:	Surfaced: 256 tonnes (252 tons) Submerged: 310 tonnes (305 tons)
Dimensions:	45m x 4.2m x 3m (148ft 2in x 13ft 9in x 9ft 10in)
Armament:	Two 450mm (17.7in) torpedo tubes
Powerplant:	Twin screw diesels, electric motors
Surface range:	Not known
Performance:	Surfaced: 12 knots Submerged: 8 knots

Flutto

Flutto was one of a class of medium submarines planned in three groups, all to be completed by the end of 1944. In the end only eight of the first group were finished. Two of the class, *Grongo* and *Merena*, were each fitted with four cylinders for the transport of human torpedoes. Of those that did become operational, *Tritone* was sunk off Bougie on 19 January 1943 by gunfire from the British destroyer *Antelope* and the Canadian corvette *Port Arthur; Gorgo* was sunk off the Algerian coast by the American destroyer USS *Nields* on 21 May 1943; and *Flutto* was sunk off the coast of Sicily on 11 July 1943 by the British motor torpedo boats 640, 651 and 670. *Nautilo,* sunk in an air raid and refloated, was assigned to the Yugoslav Navy and renamed *Sava;* and *Marea* went to Russia as the *Z13.*

Country:	Italy
Launch date:	November 1942
Crew:	50
Displacement:	Surfaced: 973 tonnes (958 tons) Submerged: 1189 tonnes (1170 tons)
Dimensions:	63.2m x 7m x 4.9m (207ft x 23ft x 16ft)
Armament:	Six 533mm (21in) torpedo tubes, one 100mm (3.9in) gun
Powerplant:	Twin screw diesel engines, electric motors
Surface range:	10,000km (5400nm) at 8 knots
Performance:	Surfaced: 16 knots Submerged: 7 knots

Foca

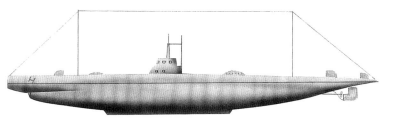

F*oca* was the only Italian submarine to have three shafts, driven by three sets of FIAT petrol engines. On 26 April 1909, in Naples harbour, an internal petrol explosion set fire to her fuel and she was scuttled to prevent the blaze spreading. She was later raised, repaired, and the central motor with its shaft and propeller were removed. The accident underlined the dangers inherent in using petrol engines in submersible craft, whose interiors quickly filled with volatile vapour that could easily be ignited by a chance spark, with disastrous consequences. After this accident the Italian Navy gave up building petrol-powered submarines. The early British boats had suffered similar problems, leading to the rapid adoption of diesel engines as the primary power source. *Foca* was finally discarded in September 1918 as World War I drew to a close.

Country:	Italy
Launch date:	September 1908
Crew:	2 + 15
Displacement:	Surfaced: 188 tonnes (185 tons)
	Submerged: 284 tonnes (280 tons)
Dimensions:	42.5m x 4.3m x 2.6m (139ft 5in x 14ft x 8ft 7in)
Armament:	Two 450mm (17.7in) torpedo tubes
Powerplant:	Petrol engines; electric motors
Surface range:	Not known
Performance:	Surfaced: Not known
	Submerged: Not known

Foca

Foca was one of three minelaying submarines built for the Italian Navy just before World War II, the others being *Atropo* and *Zoea*. As first completed, their 100mm (3.9in) gun was mounted in a training turret, in the after part of the conning tower. This gun was later removed and mounted in the traditional deck position, forward of the conning tower. Torpedo armament was sacrificed to provide two mine chutes at the stern. The class leader, *Foca*, was lost on 15 October 1940 while laying a mine barrage off Haifa, Palestine; it was thought that she had probably run into a British minefield. *Atropo* and *Zoea* survived the war and were discarded in 1947. Late in 1943, *Atropo* was used by the Allies to run supplies to British garrisons on the Aegean islands of Samos and Leros; the Aegean was heavily patrolled by enemy MTBs, and this was the safest method.

Country:	Italy
Launch date:	26 June 1937
Crew:	60
Displacement:	Surfaced: 1354 tonnes (1333 tons) Submerged: 1685 tonnes (1659 tons)
Dimensions:	82.8m x 7.2m x 5.3m (271ft 8in x 23ft 6in x 17ft 5in)
Armament:	Six 533mm (21in) torpedo tubes; one 100mm (3.9in) gun
Powerplant:	Twin screw diesel engines, electric motors
Surface range:	15,742km (8500nm) at 8 knots
Performance:	Surfaced: 15.2 knots Submerged: 7.4 knots

Foxtrot class

Built in the periods 1958–68 (45 units) and 1971–74 (17 units), the Foxtrot-class diesel-electric submarine remained in production at a slow rate for export, the last unit being launched in 1984. The class proved to be the most successful of the post-war Russian conventional submarine designs, 62 serving with the former Soviet Navy. Three Soviet Navy Fleet Areas operated Foxtrot, and the Mediterranean and Indian Ocean Squadrons regularly had these boats deployed to them. The Foxtrots were used more regularly for long-range ocean patrols than Russia's SSNs. The first foreign recipient of the type was India, which received eight new boats between 1968 and 1976. India was followed by Libya, with six units received between 1976 and 1983, and Cuba, three boats being handed over between 1979 and 1984. All Russian Foxtrots were withdrawn by the late 1980s.

Country:	Russia
Launch date:	1959 (first unit)
Crew:	80
Displacement:	Surfaced: 1950 tonnes (981 tons) Submerged: 2500 tonnes (2540 tons)
Dimensions:	91.5m x 8m x 6.1m (300ft 2in x 26ft 3in x 20ft)
Armament:	Ten 533mm (21in) torpedo tubes
Powerplant:	Three shafts, three diesel engines and three electric motors
Surface range:	10,190km (5500nm) at 8 knots
Performance:	Surfaced: 18 knots Submerged: 16 knots

Francesco Rismondo

Francesco Rismondo was the former Yugoslav submarine *Ostvenik* (N1), captured on 17 April 1941 at Cattaro by the Italian Navy. She was one of two Yugoslav boats built by Ateliers et Chantiers de la Loire at Nantes; the other was the *Smeli* (N2), which was captured at the same time. A third boat, also taken by the Italians at Cattaro, was the Vickers-built *Hrabri* (N3), whose sister ship *Nebojsa* succeeded in getting away before the Italians occupied Dalmatia. *Francesco Rismondo* was captured by German forces at Bonifacio on 14 September 1943, following the Italian armistice with the Allies, and sunk by them at the same port four days later. After the war the Italian submarine *Nautilo* was awarded to the Yugoslav Navy as war booty, serving as the *Sava* until 1970.

Country:	Italy
Launch date:	14 February 1929
Crew:	45
Displacement:	Surfaced: 676 tonnes (665 tons)
	Submerged: 835 tonnes (822 tons)
Dimensions:	66.5m x 5.4m x 3.8m (218ft 2in x17ft 9in x 12ft 4in)
Armament:	Six 551mm (21.7in) torpedo tubes; one 100mm (3.9in) gun
Powerplant:	Two diesel engines, two electric motors
Surface range:	5003km (2700nm) at 10 knots
Performance:	Surfaced: 14.5 knots
	Submerged: 9.2 knots

Fratelli Bandiera

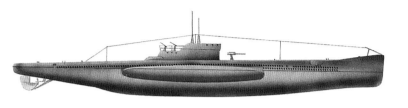

Fratelli Bandiera was leader of a class of four, the others being the *Luciano Manara, Ciro Menotti* and *Santorre Santarosa*. The latter was torpedoed by the British MTB 260 on 20 January 1943, having run aground off Tripoli; she was eventually scuttled. Initially, the boats of this class had a maximum speed of 17.9 knots on the surface and just over nine knots when submerged. They tended to plunge into oncoming waves, however, and this poor stability made it necessary to mount bulges on either side of the hull, creating extra drag. The engineer responsible for this class was Curio Bernardis. *Bandiera, Manara* and *Menotti* were used for training and transport duties in World War II, undergoing some modifications that included the raising of the bow and a reduction in size of the conning tower. All were discarded in 1948.

Country:	Italy
Launch date:	7 August 1929
Crew:	52
Displacement:	Surfaced: 880 tonnes (866 tons)
	Submerged: 1114 tonnes (1096 tons)
Dimensions:	69.8m x 7.2m x 5.2m (229ft x 23ft 8in x 17ft)
Armament:	Eight 533mm (21in) torpedo tubes; one 100mm (3.9in) gun
Powerplant:	Two sets diesel engines, two electric motors
Surface range:	8797km (4750nm) at 8.5 knots
Performance:	Surfaced: 15.1 knots
	Submerged: 8.2 knots

Frimaire

Frimaire was one of the Brumaire class of 16 submarines launched in 1911–13. All 16 boats operated in the Mediterranean during World War I. One of them, *Bernouilli*, infiltrated into Cattaro harbour on 4 April 1916 and torpedoed the Austrian destroyer *Csepel,* blowing off her stern, and another, *Le Verrier,* accidentally rammed the German *U47* on 28 July 1918 after an unsuccessful torpedo engagement. Three were lost; *Fourcault* was sunk off Cattaro by Austrian aircraft, *Curie* was captured at Pola after becoming trapped in the harbour and recommissioned by the Austrians as the *U14* (she was recovered by the French at the end of the war) and *Joule* was mined in the Dardanelles. *Frimaire* was stricken from the navy list in 1923. The boats of this class were all named after the months of the French Revolutionary calendar.

Country:	France
Launched:	26 August 1911
Crew:	29
Displacement:	Surfaced: 403 tonnes (397 tons)
	Submerged: 560 tonnes (551 tons)
Dimensions:	52.1m x 5.14m x 3.1m (170ft 11in x 17ft 9in x 10ft 2in)
Armament:	Six 450mm (17.7in) torpedo tubes
Powerplant:	Two-shaft diesel engines, electric motors
Surface range:	3150km (1700nm) at 10knots
Performance:	Surfaced: 13 knots
	Submerged: 8 knots

Fulton

Fulton was laid down at Cherbourg in late 1913, but was not completed until July 1920 because other types of warship were allocated a higher priority in the French naval construction programme. *Fulton* was originally designed with two 2000hp turbines, altered to diesel machinery in the course of construction. The submarine was named after the American Robert Fulton, designer of France's first submersible, the *Nautilus,* which was launched in 1800. An ardent pacifist, Fulton's desire was to build submarines to destroy the world's battle fleets; but the submarine was then far too crude to influence naval warfare and was doomed to remain little more than a toy for a century, more lethal to its operator than the enemy. It took another American, John Holland, to design a practical submersible.

Country:	France
Launch date:	April 1919
Crew:	45
Displacement:	Surfaced: 884 tonnes (870 tons) Submerged: 1267 tonnes (1247 tons)
Dimensions:	74m x 6.4m x 3.6m (242ft 9in x 21ft x 11ft 10in)
Armament:	Eight 450mm (17.7in) torpedo tubes; two 75mm (3in) guns
Powerplant:	Twin screw, diesel engines; electric motors
Surface range:	7964km (4300nm) at 10 knots
Performance:	Surfaced: Not known Submerged: Not known

G1

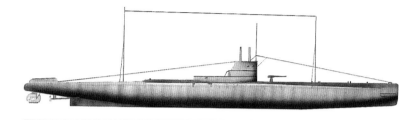

The G class of 14 boats, based on the E-class design, was ordered by the British Admiralty in 1914 in response to information that Germany was about to build a fleet of double-hulled, oceangoing submarines. Two of the G-class boats were lost in action during World War I and two more through accidental causes. One of their main tasks during the war was to ambush U-boats trying to pass through the Channel. The boats had an unusual armament arrangement in that they were fitted with torpedo tubes of different calibres: one 533mm (21in) and four 457mm (18in), the larger-calibre weapon being intended for use against armoured targets. One G-class boat, *G7*, had the unhappy distinction of being the last British submarine lost in World War I, failing to return from a North Sea patrol on 1 November 1918.

Country:	Britain
Launch date:	August 1915
Crew:	31
Displacement:	Surfaced: 704 tonnes (693 tons) Submerged: 850 tonnes (836 tons)
Dimensions:	57m x 6.9m x 4.1m (187ft x 22ft 13ft 6in)
Armament:	Four 457mm (18in) torpedo tubes, one 533mm (21in) torpedo tube, one 76mm (3in) gun
Powerplant:	Twin screw diesel-electric motors
Surface range:	4445km (2400nm) at 12.5 knots
Performance:	Surfaced: 14.25 knots Submerged: 9 knots

Gal

G*al* was one of three German-designed Type 206 boats built by Vickers of Barrow in the UK in the mid-1970s, following a contract signed in April 1972. *Gal,* laid down in 1973 and commissioned in December 1976, ran aground on her delivery voyage, but was repaired. The other two boats, *Tanin* and *Rahav,* were commissioned in June and December 1977 respectively. The Type 206 was a development of the Type 205; built of high tensile non-magnetic steel, it was intended for coastal operations and had to conform with treaty limitations on the maximum tonnage allowed for West Germany. New safety devices for the crew were fitted, and the weapons fit allowed for the carriage of wire-guided torpedoes. The Type 206 was just one of a range of similar boats offered for export. The Gal class were all decommissioned by 2002.

Country:	Israel
Launch date:	2 December 1975
Crew:	22
Displacement:	Surfaced: 427 tonnes (420 tons) Submerged: 610 tonnes (600 tons)
Dimensions:	45m x 4.7m x 3.7m (147ft 8in x 15ft 5in x 12ft 2in)
Armament:	Eight 533mm (21in) torpedo tubes
Powerplant:	Single shaft, two diesels, one electric motor
Surface range:	7038km (3800nm) at 10 knots
Performance:	Surfaced: 11 knots Submerged: 17 knots

Galatea

Galatea was one of 12 boats of the Sirena class, which was an improvement on the basic '600' design with improved seakeeping qualities, higher speed and better handling when submerged. Many of the 12 boats underwent modifications during their careers, and during the Spanish civil war Sirena-class boats carried out 18 extended patrols. The boats could carry a total of 12 533mm (21in) torpedoes, and their AA armament was gradually upgraded as the Second World War progressed. From 1940-43 they saw action in the Mediterranean, and all but *Galatea* were war losses. One of them, *Topazio,* was sunk in error by RAF aircraft south-east of Sardinia four days after Italy concluded an armistice with the Allies, the cause being lack of identification signals from the submarine. *Galatea* was discarded in February 1948.

Country:	Italy
Launch date:	5 October 1933
Crew:	45
Displacement:	Surfaced: 690 tonnes (679 tons) Submerged: 775 tonnes (701 tons)
Dimensions:	60.2m x 6.5m x 4.6m (197ft 5in x 21ft 2in x 15ft)
Armament:	Six 533mm (21in) torpedo tubes; one 100mm (3.9in) gun
Powerplant:	Twin screw diesel/electric motors
Surface range:	9260km (5000nm) at 8 knots
Performance:	Surfaced: 14 knots Submerged: 7.7 knots

Galathée

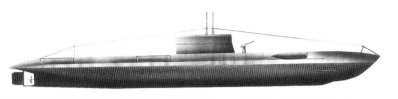

On the outbreak of World War II, the group of medium-range submarines to which *Galathée* belonged was the largest class of such vessels in the French Navy, and they operated intensively until the French collapse in June 1940. *Galathée* was one of a batch of three boats built by Ateliers Loire-Simonot; they were laid down in 1923 and completed in 1927. In spite of having a complex torpedo layout, involving a double revolving mounting, *Galathée* and her consorts were successful ships. She was scuttled at Toulon on 27 November 1942, shortly before the harbour was occupied by troops of the II SS Armoured Corps; this action followed the Allied landings in North Africa earlier in the month. Most of the boats that remained under the control of Vichy France lay idle from June 1940 to the end of 1942.

Country:	France
Launch date:	18 December 1925
Crew:	41
Displacement:	Surfaced: 619 tonnes (609 tons) Submerged: 769 tonnes (757 tons)
Dimensions:	64m x 5.2m x 4.3m (210ft x 17ft x 14ft)
Armament:	Seven 551mm (21.7in) torpedo tubes; one 76mm (3in) gun
Powerplant:	Twin screw diesel/electric motors
Surface range:	6485km (3500nm) at 7.5 knots
Performance:	Surfaced: 13.5 knots Submerged: 7.5 knots

Galerna

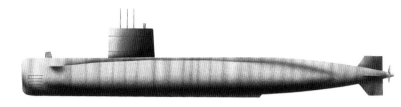

G<i>alerna</i> is a medium-range submarine built to the design of the French Agosta class. She marked a major step forward in Spanish submarine technology. This submarine and her three sisters could carry 16 reload torpedoes or nine torpedoes and 19 mines. A full sonar kit is carried, comprising one active and one passive set. The first two boats, *Galerna* and *Siroco*, were ordered in May 1975, and a second pair *(Mistral* and *Tramontana*) in June 1977. The Spanish Agosta-class boats are armed with four bow torpedo tubes which are fitted with a rapid-reload pneumatic ramming system that can launch the weapons quickly but with a minimum of noise. The boats have a diving depth of 350m (1148ft). Only *Galerna* remains in Spanish naval service in 2024, as the class is finally being replaced by the delayed four-boat Isaac Peral class.

Country:	Spain
Launch date:	5 December 1981
Crew:	54
Displacement:	Surfaced: 1473 tonnes (1450 tons) Submerged: 1753 tonnes (1725 tons)
Dimensions:	67.6m x 6.8m x 5.4m (221ft 9in x 22ft4in x 17ft 9in)
Armament:	Four 551mm (21.7in) torpedo tubes
Powerplant:	Single screw diesel/electric motors
Surface range:	13,672km (7378nm) at 9 knots
Performance:	Surfaced: 12 knots Submerged: 20 knots

Galilei

The two Archimede class submarines *Galilei* and *Ferraris* both served in the Spanish Civil War, and when Italy entered World War II they were both in the Red Sea at Massawa. On the outbreak of hostilities *Ferraris* set up a patrol area off Djibouti and *Galilei* an area off Aden. On 16 June 1940, *Galilei* sank the Norwegian tanker *James Stove*. Two days later she stopped the Yugoslav steamer *Drava* but had to release her; on the following day she was sighted by the armed British anti-submarine trawler *Moonstone* and a gun battle ensured in which nearly all her officers were killed and the crew, still below, were poisoned by emissions from the air-conditioning system. She was captured and impressed into British service as the *P711*. She was used as a training boat in the East Indies and Mediterranean, and was scrapped in 1946.

Country:	Italy
Launch date:	9 March 1934
Crew:	55
Displacement:	Surfaced: 1001 tonnes (985 tons)
	Submerged: 1279 tonnes (1259 tons)
Dimensions:	70.5m x 6.8m x 4.1m (231ft 4in x 22ft 4in x 13ft 5in)
Armament:	Eight 533mm (21in) torpedo tubes; two 100mm (3.9in) guns
Powerplant:	Twin screw diesel/electric motors
Surface range:	6670km (3600nm) at 10 knots
Performance:	Surfaced: 17 knots
	Submerged: 8.5 knots

Galvani

Galvani was one of three Brin-class submarines. On 10 June 1940, at the point of Italy's entry into the war, she was in the Red Sea under Cdr Spano, and with the commencement of hostilities she set up a patrol area in the Gulf of Oman, where she sank an Indian sloop, HMIS *Pathan*. Her fate was sealed, however, by the capture of the *Galilei*, whose documents revealed the location of all Indian submarines in the Red Sea area. On 24 June she was located by British warships, which opened fire on her; she was finished off by depth charges from the sloop HMS *Falmouth*. Her sister ship and third of the class, *Guglielmotti*, was sunk off Calabria on 17 March 1942 by torpedoes from HM submarine *Unbeaten*. The class leader, *Brin*, went on to see much service as a training boat in the Indian ocean after Italy's armistice with the Allies.

Country:	Italy
Launch date:	22 May 1938
Crew:	58
Displacement:	Surfaced: 1032 tonnes (1016 tons) Submerged: 1286 tonnes (1266 tons)
Dimensions:	72.4m x 6.9m x 4.5m (237ft 6in x 22ft 14ft 11in)
Armament:	Eight 533mm (21in) torpedo tubes; one 100mm (3.9in) gun
Powerplant:	Twin screw diesel/electric motors
Surface range:	19,446km (10,500nm) at 8 knots
Performance:	Surfaced: 17.3 knots Submerged: 8 knots

Gemma

Laid down in September 1935 and completed in July 1936, *Gemma* was one of a class of ten short-range boats of the Perla class, derived in turn from the Sirena series that had been completed in 1933–34, but with a slight increase in displacement and more modern equipment. Project designer for this class was General Engineer Curio Bernardis. Their maximum diving depth was 70–80m (230–260ft). They undertook patrols on behalf of the Nationalists in the Spanish Civil War, and two were ceded to Spain for several months. Five were lost in World War II, including *Gemma*. Her early war patrol station was off Sollum in the eastern Mediterranean. She was later transferred to the south-eastern approaches to the Aegean, where, on 6 October 1940, she was sunk in error by the Italian submarine *Tricheco*.

Country:	Italy
Launch date:	21 May 1936
Crew:	45
Displacement:	Surfaced: 711 tonnes (700 tons) Submerged: 711 tonnes (830 tons)
Dimensions:	60.2m x 6.5m x 4.6m (197ft 5in x 21ft 2in x 15ft)
Armament:	Six 533mm (21in) torpedo tubes; one 100mm (3.9in) gun
Powerplant:	Twin screw diesel/electric motors
Surface range:	9260km (5000nm) at 8 knots
Performance:	Surfaced: 14 knots Submerged: 7.5 knots

General Mola

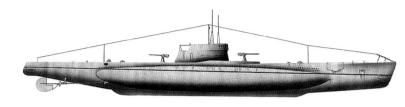

General Mola was formerly the Italian submarine *Torricelli,* which was transferred to Spain in 1937 together with the *Archimede.* To cover up the deal, the Italians built two more submarines in conditions of the strictest secrecy and gave them the same names. The two vessels remained on the Spanish Navy's inventory until the early 1950s, when they were replaced by two locally-built D-class submarines. *General Mola* bore the pennant number C5 from 1950 until she was stricken in 1959. As a matter of interest, Spain also used an ex-German Type VII U-boat until the 1960s; she was the *U573*, damaged by a Hudson aircraft of the RAF east of Gibraltar on 1 May 1942 and forced to seek refuge in a Spanish port, where she was interned. She was purchased from Germany in the following year and given the number *G7*.

Country:	Spain
Launch date:	April 1934
Crew:	55
Displacement:	Surfaced: 1001 tonnes (985 tons) Submerged: 1279 tonnes (1259 tons)
Dimensions:	70.5m x 6.8m x 4.1m (231ft 4in x 22ft 4in x 13ft 5in)
Armament:	Eight 533mm (21in) torpedo tubes; two 100mm (3.9in) guns
Powerplant:	Twin screw diesel/electric motors
Surface range:	6670km (3600nm) at 10 knots
Performance:	Surfaced: 17 knots Submerged: 8.5 knots

George Washington

In 1955, the Soviet Union began modifying six existing diesel submarines to carry nuclear-tipped ballistic missiles. America was simultaneously developing the Jupiter missile, which was to equip a projected 10,160-tonne (10,000-ton) nuclear submarine. Jupiter used a mix of highly volatile liquids for its propellant, and was posing immense problems of safety and operation. The smaller, lighter Polaris A1 presented a more suitable alternative. The nuclear submarine *Scorpion*, then building, was chosen as the delivery platform for the new weapon and a new 40m (13ft) hull section was added just aft of the conning tower to house 16 missiles in vertical launch tubes. Renamed *George Washington*, she was the first of a new type of weapons platform, and put the USA far ahead in the nuclear arms race. The five boat class was decommissioned in the 1980s.

Country:	USA
Launch date:	June 1959
Crew:	112
Displacement:	Surfaced: 6115 tonnes (6019 tons) Submerged: 6998 tonnes (6888 tons)
Dimensions:	116.3m x 10m x 8.8m (381ft 7in x 33ft x 28ft 10in)
Armament:	Sixteen Polaris missiles, six 533mm (21in) torpedo tubes
Powerplant:	Single screw, one pressurized water-cooled reactor, turbines
Range:	Unlimited
Performance:	Surfaced: 20 knots Submerged: 30.5 knots

George Washington Carver

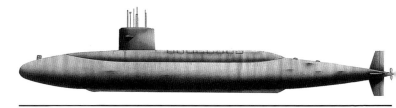

George Washington Carver was one of nine vessels of the Lafayette class, enlarged versions of the Ethan Allen class, and all were refitted with Poseidon missiles. She was laid down in April 1964 and was completed in August 1966. This class of submarine could dive to depths of up to 300m (985ft) and the nuclear reactor core provided enough energy to propel the vessel for 760,000km (347,200nm), which to all intents and purposes gave it an unlimited endurance. Like all American SSBNs, George Washington Carver had two crews which carried out alternate 68-day patrols, with 32-day refit periods between patrols. The vessels underwent an extensive refit, which took nearly two years to complete, every six years in rotation. George Washington Carver was deactivated on 2 November 1992.

Country:	USA
Launch date:	14 August 1965
Crew:	140
Displacement:	Surfaced: 7366 tonnes (7250 tons) Submerged: 8382 tonnes (8250 to6
Dimensions:	129.5m x 10m x 9.6m (424ft 10in x 32ft 10in x 31ft 10in)
Armament:	Sixteen Trident C4 missiles, four 533mm (21in) torpedo tubes
Powerplant:	Single screw, one pressurized water-cooled nuclear reactor
Range:	Unlimited
Performance:	Surfaced: 20 knots Submerged: 30 knots

Georgia

Georgia was built as one of 18 Ohio-class ballistic-missile submarines (SSBNs), which provide the principal US strategic deterrent following the drastic reduction of land- and air-based systems since 1991. Trident missile accuracy is as good as ground-based systems, and the warhead (each missile carries up to 12) has a 50 per cent higher yield than the most lethal ICBM. In 2002 conversion to SSGNs began of the oldest four boats, *Ohio*, *Michigan*, *Florida*, with *Georgia* the last to reenter service in 2008. Equipped with up to 154 Tomahawk SLCMs or 66 special forces personnel, each vessel cost around $1bn to refit. The remaining 14 SSBN boats had four launch tubes deactivated in 2017 to comply with the New START treaty. They are due to be replaced by the new Columbia class SSBNs beginning in the early 2030s.

Country:	USA
Launch date:	6 November 1982
Crew:	155
Displacement:	Surfaced: 16,866 tonnes (16,600 tons) Submerged: 19,051 tonnes (18,750 tons)
Dimensions:	170.7m x 12.8m x 10.8m (560ft x 42ft x 35ft 5in)
Armament:	Twenty-four Trident C4 missiles; four 533mm (21in) torpedo tubes
Powerplant:	Single screw, one pressurized water-cooled reactor
Range:	Unlimited
Performance:	Surfaced: 20 knots Submerged: 24 knots

Giacinto Pullino

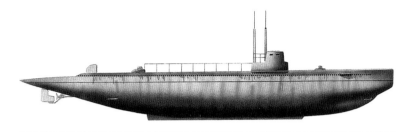

Giacinto Pullino was laid down at La Spezia dockyard in June 1912 and was completed in December 1913. During World War I she served in the Adriatic, along with most other Italian submarines of the time, where her main function was to carry out frequent reconnaissance missions along the Dalmatian coast, which was held by the Austro-Hungarians and whose many harbours and inlets provided refuges for their warships. Because of the relatively shallow waters of the Adriatic's east coast it was difficult and dangerous work, and it was during one such mission in July 1916 that she ran aground on Galiola Island, Quarnaro, and was seized by Austrian forces. She sank while being towed to Pola on 1 August 1917. In 1931 the Italian Navy raised her, but she was scrapped later the same year.

Country:	Italy
Launch date:	July 1913
Crew:	40
Displacement:	Surfaced: 350 tonnes (345 tons) Submerged: 411 tonnes (405 tons)
Dimensions:	42.2m x 4m x 3.7m (138ft 6in x 13ft 1in x 12ft 4in)
Armament:	Six 450mm (17.7in) torpedo tubes; one 57mm (2.25in) and one 47mm (1.85in) guns
Powerplant:	Twin screw diesel/electric motors
Surface range:	Not known
Performance:	Surfaced: 14 knots Submerged: 9 knots

Giacomo Nani

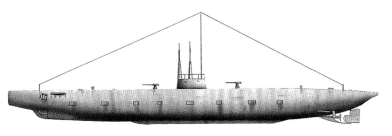

Giacomo Nani and her three sisters were fast, medium-sized submarines designed by Laurenti and Cavallini. *Giacomo Nani* was laid down in 1915, but was not completed in time to see service in World War I, where her superior speed, both surfaced and dived, would certainly have made her a formidable opponent. She was stricken from the Navy list in 1935, by which time the Italian Navy had several new classes of submarine on its inventory. Not all of them were as well designed as *Giacomo Nani;* the early pre-war classes of submarine in Italian service had numerous problems, mostly concerned with underwater handling, and were constantly modified to improve matters. Much of the design expertise built up during World War I had been lost when shipyards came under state control after 1925.

Country:	Italy
Launch date:	September 1918
Crew:	35
Displacement:	Surfaced: 774 tonnes (762 tons)
	Submerged: 938 tonnes (924 tons)
Dimensions:	67m x 5.9m x 3.8m (220ft x 19ft 4in x 12ft 6in)
Armament:	Six 450mm (17.7in) torpedo tubes; two 76mm (3in) guns
Powerplant:	Twin screw diesel/electric motors
Surface range:	Not known
Performance:	Surfaced: 16 knots
	Submerged: 10 knots

Giovanni Bausan

Giovanni Bausan was one of four submarines of the Pisani class laid down in 1925-26. This class of submarine was a joint design project between Colonel Engineer Curio Bernardis and Major Engineer Tizzoni. She was a short-range boat with an internal double hull. Because of stability problems revealed during trials, all four boats in the class were fitted with external bulges which reduced their speed by about two knots on the surface and one knot submerged. In 1940 *Giovanni Bausan* became a training ship; she was laid up in 1942 and used as a floating oil depot under the number *GR251*. Of her sister vessels, *Marcantonio Colonna* was laid up in April 1942 and broken up in 1943; *Des Geneys* was also laid up in the same month and converted to a hull for charging batteries; and *Vittorio Pisani* was laid up in March 1947.

Country:	Italy
Launch date:	24 March 1928
Crew:	48
Displacement:	Surfaced: 894 tonnes (880 tons) Submerged: 1075 tonnes (1058 tons)
Dimensions:	68.2m x 6m x 4.9m (223ft 9in x 20ft x 16ft 2in)
Armament:	Six 533mm (21in) torpedo tubes; one 120mm (4in) gun
Powerplant:	Twin screws, diesel/electric motors
Surface range:	9260km (5000nm) at 8 knots
Performance:	Surfaced:15 knots Submerged: 8.2 knots

Giovanni da Procida

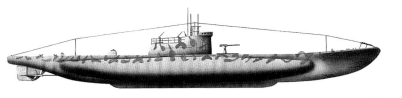

Giovanni da Procida was one of four medium sized submarines of the Mameli class, all of which survived World War II except *Pier Capponi,* which was torpedoed by HM submarine *Rorqual* south of Stromboli on 31 March 1941. Built at the Tosi yards, the boats of this class each carried 10 torpedoes. All four served on patrol duty during the Spanish Civil War. In June 1940 *da Procida* was one of the Italian submarines that operated unsuccessfully against French shipping evacuating personnel and material from southern France to North Africa, and in August she was switched to the eastern Mediterranean, patrolling off Palestine and Cyprus. She was re-engined in 1942 and took no further part in the Mediterranean war, being relegated to training duties. She was scrapped in 1948.

Country:	Italy
Launch date:	1 April 1928
Crew:	49
Displacement:	Surfaced: 843 tonnes (830 tons) Submerged: 1026 tonnes (1010 tons)
Dimensions:	64.6m x 6.5m x 4.3m (212ft x 21ft 4in x 14ft)
Armament:	Six 533mm (21in) torpedo tubes; one 102mm (4in) gun
Powerplant:	Twin shafts, diesel engines, electric motors
Surface range:	5930km (3200nm) at 10 knots
Performance:	Surfaced: 17 knots Submerged: 7 knots

Giuliano Prini

The *Giuliano Prini* is one of four SSKs laid down by Fincantieri, Monfalcone, for the Italian Navy between 1984 and 1992. The first two boats, *Salvatore Pelosi* and *Giuliano Prini,* were ordered in March 1983 and the second pair, *Primo Longobardo* and *Gianfranco Gazzana Priaroggia,* in July 1988. The latter two boats have a slightly longer hull to provide space for the installation of SSMs. Test diving depth of the class is 300m (985ft) and the hull structure can withstand pressures down to 600m (1970ft) before crushing. The boats have an operational endurance of 45 days and a submerged range of 217nm at 4 knots. The SSK is a vital component of Italy's naval inventory because of the need to protect the country's long coastline from infiltration. In the early 2000s all four boats had a refit. The class are due to be replaced by Type 212A boats in the late 2020s.

Country:	Italy
Launch date:	12 December 1987
Crew:	50
Displacement:	Surfaced: 1500 tonnes (1476 tons)
	Submerged: 1689 tonnes (1662 tons)
Dimensions:	64.4m x 6.8m x 5.6m (211ft 2in x 22ft 3in x 18ft 5in)
Armament:	Six 533mm (21in) torpedo tubes
Powerplant:	Single shaft, diesel/electric motors
Surface range:	17, 692km (9548nm) at 11 knots
Performance:	Surfaced: 11 knots
	Submerged: 19 knots

Giuseppe Finzi

The *Giuseppe Finzi* was one of four boats of the Calvi class, whose class leader, *Pietro Calvi*, enjoyed considerable success against Allied convoys in the Atlantic before being scuttled after a violent battle with the sloop HMS *Lulworth* and other warships on 15 July 1942. Most Italian submarines in the Atlantic operated in the Azores area. *Giuseppi Finzi* was one of the first Italian submarines to be deployed to the Atlantic, but it was not until 1942 that she began to register some sinkings. At this time she was commanded by Cdr Giudice and was operating in the South Atlantic as part of the Da Vinci group of submarines (*Finzi, Torelli, Tazzoli* and *Morosini)* which was probably the most successful in the Italian Navy. On 9 September 1943 she was captured by the Germans at Bordeaux and numbered *UIT21*; she was scuttled on 25 August 1944.

Country:	Italy
Launch date:	29 June 1935
Crew:	77
Displacement:	Surfaced: 1574 tonnes (1550 tons) Submerged: 2093 tonnes (2060 tons)
Dimensions:	98.3m x 9.1m x 5.3m (322ft 4in x 30ft x 17ft 4in)
Armament:	Eight 533mm (21in) torpedo tubes; two 120mm (4.7in) guns
Powerplant:	Twin screw, diesel/electric motors
Surface range:	24,817km (13,400nm) at 8 knots
Performance:	Surfaced: 16.8 knots Submerged: 4.7 knots

Glauco

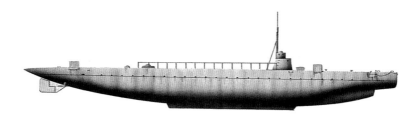

Built at the Venice Naval Dockyard to designs by Engineer Lurenti, *Glauco* belonged to the first mass-produced group of submarines built for the Italian Navy. When *Glauco* was laid down in 1903, petrol engines were still being fitted to submarines in spite of the volatile nature of the fuel they used, which led to numerous accidents. *Glauco's* engines developed 600hp, giving a surface range of 1710km (922nm) at eight knots. Submerged, her electric motors developed 170hp and range was 65km (35nm) at 5 knots. There were some differences between the boats, for example the torpedo tubes, three for *Glauco* reduced to two in the rest. During the war these boats were employed for harbour defence at Brindisi and Venice. *Glauco* was removed from the effective list in 1916, having been used for training for some time.

Country:	Italy
Launch date:	July 1905
Crew:	30
Displacement:	Surfaced: 160 tonnes (157 tons)
	Submerged: 243 tons (240 tons)
Dimensions:	36.8m x 4.3m x 2.6m (120ft 9in x 14ft x 8ft 6in)
Armament:	Three 450mm (17.7in) torpedo tubes
Powerplant:	Twin screws, petrol engines, electric motors
Surface range:	1710km (922nm) at 8 knots
Performance:	Surfaced: 14 knots
	Submerged: 7 knots

Glauco

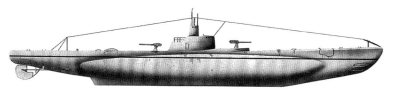

Glauco and her sister boat *Otaria* were originally ordered for the Portuguese Navy under the names *Delfim* and *Espadarte,* so they were completed for service in the Italian Navy. *Glauco's* first war station in June 1940 was off the Algerian coast, and in September she was transferred to the Atlantic, her very long range making her well suited for operations in that theatre. On 27 June 1941 she was scuttled west of Gibraltar after having been damaged by gunfire from the destroyer HMS *Wishart.* Her sister boat, *Otaria,* was used to transport fuel and supplies to Axis forces in Tunisia early in 1943, this being at a time when Allied air power and naval forces combined were wreaking terrible havoc on Axis upply convoys in transit across the Mediterranean. She survived the war and was discarded in 1948.

Country:	Italy
Launch date:	5 January 1935
Crew:	59
Displacement:	Surfaced: 1071 tonnes (1055 tons) Submerged: 1346 tonnes (1325 tons)
Dimensions:	73m x 7.2m x 5m (239ft 6in x 23ft 8in x 16ft 6in)
Armament:	Eight 533mm (21in) torpedo tubes; two 100mm (3.9in) guns
Powerplant:	Twin screws, diesel/electric motors
Surface range:	10,000km (5390nm) at 8 knots
Performance:	Surfaced: 17.3 knots Submerged: 8.6 knots

Golf I

By the 1950s, Russia had embarked upon a massive submarine programme that would initially give her a larger fleet of submarines than any other country. Twenty-three Golf I-class boats were completed between 1958 and 1962, and entered service at a rate of six or seven a year. One unit was built in China from Russian-supplied components. The ballistic missiles were housed vertically in the rear section of the extended fin, which produced a great deal of resistance underwater and reduced speed, as well as generating high noise levels; however, the boats could be driven by a creep motor, giving quiet operation and very long endurance. Thirteen Golf I boats were modified to Golf II standard starting in 1965, using the SS-N-5 ballistic missile. Code-named Sark by NATO, this was a single-stage, liquid-fuelled missile with a range of 1400km (750nm).

Country:	Russia
Launch date:	1977
Crew:	86
Displacement:	Surfaced: 2336 tonnes (2300 tons) Submerged: 2743 tonnes (2700 tons)
Dimensions:	100m x 8.5m x 6.6m (328ft x 27ft 11in x 21ft 8in)
Armament:	Three SS-N-4 ballistic missiles; ten 533mm (21in) torpedo tubes
Powerplant:	Triple screws, diesel/electric motors
Surface range:	36,510km (19,703nm) at 10 knots
Performance:	Surfaced: 17 knots Submerged: 14 knots

Goubet I

At the end of the nineteenth century, Great Britain was considered France's main enemy, and, since British industry was stronger, the French tried to build up a navy of small but numerous coastal combatants such as torpedo boats and submarines.The greatest difficulty facing early submarine designers was to find an acceptable form of underwater propulsion. Steam power and compressed air were being tested, but they had limitations. An answer appeared in 1859 when Plante invented the lead accumulator. By 1880, this had been improved by coating the surface with red lead. At long last, the submarine designers had access to a power source that no longer relied on oxygen to function. *Goubet I* had a pointed, cylindrical hull with an observation dome. She was one of the first successful submarines, but was discarded because of her small size.

Country:	France
Launch date:	1887
Crew:	Two
Displacement:	Surfaced: 1.6 tonnes /tons
	Submerged: 1.8 tonnes /tons
Dimensions:	5m x 1.7m x 1m (16ft 5in x 5ft 10in x 3ft 3in)
Armament:	None
Powerplant:	Single screw electric motor
Surface range:	Not known
Performance:	Surfaced: 5 knots
	Submerged: Not known

Goubet II

G*oubet II* was laid down one year after the launch of *Goubet I*. Motive power was provided by a 4hp Siemens electric road car engine, and range at full speed was about 38km, or just over 20 nautical miles. Motive power was derived from a battery of Laurent-Cely accumulators carried in the lower portion of the hull. After a series of trials, *Goubet II,* like her predecessor, was rejected because of her small size. However, she was a well-planned, successful craft and the valuable experience gained during her building and trials was put to good use by later submarine designers. The acceptable solution to the propulsion problem – a combination of diesel engines for surface travel, and electric motors for use submerged – was not far away. It would take the Germans, however, to realize its full potential.

Country:	France
Launch date:	1889
Crew:	2
Displacement:	Surfaced: 4.5 tonnes/tons Submerged: 5 tonnes/tons
Dimensions:	8m x 1.8m x 1.8m (26ft 3in x 6ft x 6ft
Armament:	None
Powerplant:	Single screw, electric motor
Surface range:	38km (20.5nm) at full speed
Performance:	Surfaced: 6 knots Submerged: Not known

Grayback

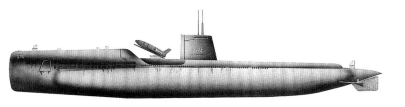

Grayback and her sister vessel, *Growler,* were originally intended to be attack submarines, but in 1956 their design was modified to provide a missile-launching capability using the Regulus, a nuclear-tipped high-altitude cruise missile which was launched by solid-fuel boosters and then guided to its target by radio command signals from the submarine, cruising at periscope depth. Both submarines were withdrawn from service in 1964, when the Regulus programme ended, but *Grayback* was subsequently converted to an Amphibious Transport Submarine (LPSS), capable of carrying 67 Marines and their assault craft. Her torpedo tubes and attack capability were retained. *Growler* was also to have been converted, but this was deferred because of rising costs. As a command ship, Grayback had a crew of 96 and could accommodate 10 officers and 75 men.

Country:	USA
Launch date:	2 July 1957
Crew:	84
Displacement:	Surfaced: 2712 tonnes (2670 tons) Submerged: 3708 tons (3650 tons)
Dimensions:	102m x 9m x (335ft x 30ft)
Armament:	Four Regulus missiles, eight 533mm (21in) torpedo tubes
Powerplant:	Twin screws, diesel/electric motors
Surface range:	14,824km (8000nm) at 10 knots
Performance:	Surfaced: 20 knots Submerged: 17 knots

Grayling

G *rayling* was formerly numbered *D2*, and later became *S18*. She was one of the last submarines in the US Navy to have petrol engines, which were a source of constant anxiety to her 15-man crew. *Grayling's* engines developed 600hp, giving her a surface range of 2356km (1270nm) at cruising speed. The three boats of this D class began service off the East coast. All American submarines were named after fish; during World War II, so many new boats were built that the Navy ran out of existing fish names, so they invented names that in the future could be given to fish of newly-discovered species. When America entered the war, over half the submarines in commission were of World War I vintage. It was a telling indictment of US Naval policy during the interwar years, and it resulted in needless losses in the early months of the Pacific War.

Country:	USA
Launch date:	June 1909
Crew:	15
Displacement:	Surfaced: 292 tonnes (288 tons) Submerged: 342 tonnes (337 tons)
Dimensions:	41m x 4.2m x 3.6m (135ft x 13ft 9in x 12ft)
Armament:	Four 457mm (18in) torpedo tubes
Powerplant:	Twin screws, two petrol engines, two electric motors
Surface range:	2356km (1270nm) at 10 knots
Performance:	Surfaced: 12 knots Submerged: 9.5 knots

Grongo

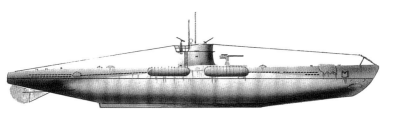

G *rongo* was one of the 12-strong Flutto class of submarines, among the last to be built for the Italian Navy before the Armistice with the Allied powers was concluded. The Fluttos were in turn developed from the Argo class, which had been launched in 1936. *Grongo's* diesel engines developed 2400hp, and surfaced range was 10,260km (5530nm) at eight knots. Her electric motors developed 800hp, and submerged range was 128km (69nm) at four knots or 13km (7nm) at seven knots. *Grongo* was scuttled at La Spezia in 1943, but the Germans raised her and commissioned her into the German Navy as the *UIT20*. She was sunk in a British air attack on Genoa on 4 September 1944. One of this class, *Marea*, was later transferred to Soviet Russia under the terms of the Peace Treaty and served until 1960 as the *Z13*.

Country:	Italy
Launch date:	6 May 1943
Crew:	50
Displacement:	Surfaced: 960 tonnes (945 tons) Submerged: 1130 tonnes (1113 tons)
Dimensions:	63m x 6.9m x 4.8m (207ft 2in x 23ft x 16ft)
Armament:	Six 533mm (21in) torpedo tubes, one 100mm (3.9in) gun
Powerplant:	Twin screws, diesel/electric motors
Surface range:	10,260km (5530nm) at 8 knots
Performance:	Surfaced: 16 knots Submerged: 7 knots

Grouper

G *rouper* was originally completed as one of the Gato class, and ten years later she was converted into one of the first hunter/killer submarines (SSK) dedicated specifically to tracking down and destroying enemy submarines. The concept required that the hunter/killer submarine be very quiet and carry long-range listening sonar with high bearing accuracy. So equipped, the submarine could lie in wait off enemy bases, or in narrow straits, and intercept the enemy boats as they moved out to patrol. *Grouper* was converted in 1951, and in 1958 she became the sonar test submarine for the Underwater Sound Laboratory at New London. The work she did in this respect was vital in building up a library of underwater 'sound signatures'. The submarine was decommissioned in 1968, and was scrapped in 1970.

Country:	USA
Launch date:	7 October 1941
Crew:	80
Displacement:	Surfaced: 1845 tonnes (1816 tons) Submerged: 2463 tonnes (2425 tons)
Dimensions:	94.8m x 8.2m x 4.5m (311ft 3in x 27ft x 15ft)
Armament:	Ten 533mm (21in) torpedo tubes
Powerplant:	Twin screws, diesel/electric motors
Surface range:	19,300km (10,416nm) at 10 knots
Performance:	Surfaced: 20.25 knots Submerged: 10 knots

Guglielmo Marconi

Guglielmo Marconi was lead vessel of a class of six submarines, all but one of which were lost during World War II. (The exception was the *Luigi Torelli*, which was captured by the Japanese at Singapore when the news of Italy's armistice with the Allies broke, and ended the war immobilized at Kobe). *Marconi* was transferred to the Atlantic in the closing months of 1940 and, under Cdr Chialamberto, scored her first success in November when she sank a merchantman, already disabled by air attack, in the central Atlantic. She was lost in November 1941; the cause was never fully established, but one theory is that she was sunk in error by the German submarine *U67* (Lt-Cdr Mäller-Stückheim), which was operating in the area at the time and which had registered some sinkings of British vessels.

Country:	Italy
Launch date:	30 July 1939
Crew:	56
Displacement:	Surfaced: 1214 tonnes (1195 tons) Submerged: 1513 tonnes (1490 tons)
Dimensions:	76.5m x 6.8m x 4.7m (251ft x 22ft 4in x 15ft 5in)
Armament:	Eight 533mm (21in) torpedo tubes; one 100mm (3.9in) gun
Powerplant:	Twin screw, diesel/electric motors
Surface range:	19,950km (10,750nm) at 8 knots
Performance:	Surfaced: 18 knots Submerged: 8.2 knots

Gustave Zédé

After overcoming some initial problems of inadequate power from the 720-cell batteries, together with their excessive weight which gave her uneven diving characteristics, *Gustave Zédé* became one of the world's first successful submarines, completing over 2500 dives without incident. During her trials she made the 66km (35nm) journey from Toulon to Marseille underwater. *Gustave Zédé* was probably the first submarine to be fitted with an effective periscope, and this innovation put France at the forefront of submarine technology. She was given a tall conning tower with a platform for surface lookouts. The hull was made up from 76 sections of Roma-bronze, and all the controls in the boat were placed centrally under the conning tower. *Gustave Zédé* was stricken from the Navy List in 1909.

Country:	France
Launch date:	June 1893
Crew:	19
Displacement:	surfaced: 265 tonnes (261 tons) submerged: 274 tonnes (270 tons)
Dimensions:	48.5m x 3.2m x 3.2m (159ft x 10ft 6in x 10ft 6in)
Armament:	One 450mm (17.7in) torpedo tube
Powerplant:	Single screw, electric motor
Surface range:	Not known
Performance:	surfaced: 9.2 knots submerged: 6.5 knots

Gustave Zédé

Gustave Zédé was one of the last steam-driven submarines built for the French Navy, and at the time of her completion in October 1914 she was one of the fastest submarines in the world. Her two reciprocating engines developed 1640hp, and her electric motors produced 1640hp. Her sister ship, Néréide, was fitted with the originally specified diesel motors which were only half as powerfulas first envisaged. *Gustave Zédé's* underwater range was 256km (138nm) at 5 knots. In 1921-22 the boat was fitted with diesel engines taken from the former German submarine *U165*. At the same time *Gustave Zédé* was fitted with a new bridge, and her fuel capacity was increased as two ballast tanks were converted to carry diesel fuel. She served in the Adriatic during the First World War, and was stricken from the Navy List in 1937.

Country:	France
Launch date:	May 1913
Crew:	32
Displacement:	Surfaced: 862 tonnes (849 tons) Submerged: 1115 tonnes (1098 tons)
Dimensions:	74m x 6m x 3.7m (242ft 9in x 19ft 8in x 12ft 2in)
Armament:	Eight 450mm (17.7in) torpedo tubes
Powerplant:	Twin screws, reciprocating engines, electric motors
Surface range:	2660km (1433nm) at 10 knots
Performance:	Surfaced: 9.2 knots Submerged: 6.5 knots

Gymnôte

Dupuy de Lôme, whose name was to become celebrated in the field of maritime design and engineering, prepared the initial drawings for *Gymnôte*, but after his death the plans were revised by Gustave Zédé, who produced a single-hull steel submarine with a detachable lead keel. Electric power was provided by 204 cells spread along the lower part of the hull. Ordered in 1886, *Gymnôte* made over 2000 dives in all. She sank in dock at Toulon in 1907, was raised, and was scrapped in the following year. *Gymnôte* and *Gustave Zédé* were the last French submersibles to depend on electric motive power alone. They had proved the concept of its use, but they had also proved that it did not provide the whole of the answer. From now on, thinking would turn increasingly to a combination of diesel and electric power.

Country:	France
Launch date:	September 1888
Crew:	5
Displacement:	Surfaced: 30 tonnes/tons Submerged: 31 tonnes/tons
Dimensions:	7.3m x 1.8m x 1.6m (58ft 5in x 6ft x 5ft 6in)
Armament:	Two 355mm (14in) torpedo tubes
Powerplant:	Single screw, electric motor
Surface range:	Not known
Performance:	Surfaced: 7.3 knots Submerged: 4.2 knots

H1

H1 was one of eight Italian submarines that were exact copies of the British H class, and were all built by the Canadian Vickers Company, Montreal. She and her sisters were unique in that their electric motors developed more power than their diesels. One of the class, *H5*, was sunk in error by the British submarine *HB1* in the South Adriatic on 16 April 1918. All the H-class boats went on to serve in the early stages of World War II, and formed a submarine group that patrolled the Gulf of Genoa in the days after Italy's entry into the war in June 1940. *H1* was armed with a 76mm (3in) gun in 1941, shortly before she was withdrawn from first-line service. The boats served generally on training duties in home waters. *H31* was lost, cause unknown, in the Bay of Biscay, and *H49* was sunk off the Dutch coast. *H1* was scrapped in 1947.

Country:	Italy
Launch date:	16 October 1916
Crew:	27
Displacement:	Surfaced: 370 tonnes (365 tons) Submerged: 481 tonnes (474 tons)
Dimensions:	45.8m x 4.6m x 3.7m (150ft 3in x 15ft 4in x 12ft 5in)
Armament:	Four 450mm (17in)torpedo tubes
Powerplant:	Twin screws, diesel/electric motors
Surface range:	Not known
Performance:	Surfaced: 12.5 knots Submerged: 8.5 knots

H4

H4 was one of 17 boats ordered for the Imperial Russian Navy under the 1915 Emergency Programme. With the collapse of Tsarist Russia, a few were seized by the Bolsheviks and commissioned. Eleven were actually delivered to Russia in sections for assembly at the Baltic Shipyards. *H4*'s contract was cancelled, however, and she was purchased for the United States Navy from her builders, the Electric Boat Company, in 1918. The boats had been built to a Holland design identical with the boats built for Britain, Italy and the USA. In 1920 *H4* was renumbered *SS147*. The US H class had a designed depth limit of 6m (20ft) and, despite engine problems, were considered successful boats. *H4* was stricken in 1930 and broken up in 1931. This class should not be confused with the British and Chilean H class, also built by the Electric Boat Company.

Country:	USA
Launch date:	October 1918
Crew:	35
Displacement:	Surfaced: 398 tonnes (392 tons) Submerged: 529 tonnes (521 tons)
Dimensions:	45.8m x 4.8m x 3.8m (150ft 3in x 15ft 9in x 12ft 6in)
Armament:	Four 457mm (18in) torpedo tubes
Powerplant:	Twin screws, diesel/electric motors
Surface range:	3800km (7041nm)
Performance:	Surfaced: 14 knots Submerged: 10 knots

Ha 201 class

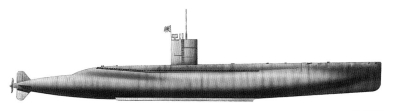

Ordered under a crash programme of 1943-44, these small submarines had a high underwater speed and excellent manoeuvrability, and were designed for the sole purpose of defending the Japanese Home Islands against American warships. Large numbers were planned, and it was hoped that the production schedule could be met by prefabricating parts of the hull in the workshops and assembling them on the slipway. Electric welding was extensively used, and the first unit, *Ha 201*, was laid down in the Sasebo Naval Yard on 1 May 1945 and completed on 31 May 1945. Owing to the critical shortage of materials and to American bombing, only 10 units had been completed by the end of the war, and none carried out any active patrols. *Ha 201* was scuttled by the US Navy in April 1946.

Country:	Japan
Launch date:	May 1945
Crew:	22
Displacement:	Surfaced: 383 tonnes (377 tons) Submerged: 447 tonnes (440 tons)
Dimensions:	50m x 3.9m x 3.4m (164ft x 13ft x 11ft 3in)
Armament:	Two 533mm (21in) torpedo tubes; one 7.7mm AA gun
Powerplant:	Single shaft diesel/electric motor
Surface range:	5559km (3000nm) at 10 knots
Performance:	Surfaced: 10.5 knots Submerged: 13 knots

Hai Lung

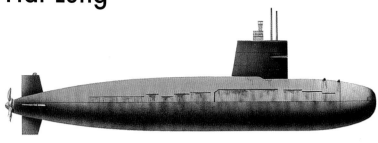

The Taiwanese Navy was set up in the 1950s to counter the threat of invasion from mainland China. Two of its most modern and effective warships are the diesel-electric submarines of the Hai Lung (Sea Dragon) class. They are modified Zwaardvis-class submarines purchased from Holland, which were probably the most efficient conventional submarine design of the 1970s. The Zwaardvis boats were themselves based on the US Barbel class, but had many differences in detail and equipment. *Hai Lung* and her sister *Hai Hu* were commissioned in December 1987, and were the first export orders for Dutch-built submarines. Their delivery was made in the face of strong protests from the People's Republic of China, and a further order for two boats was blocked by the Netherlands government. *Hai Lung* and *Hai Hu* were upgraded with Harpoon missiles in 2013.

Country:	Taiwan
Launch date:	October 1986
Crew:	67
Displacement:	Surfaced: 2414 tonnes (2376 tons) Submerged: 2702 tonnes (2660 tons)
Dimensions:	66m x 8.4m x 7.1m (216ft 6in x 27ft 7in x 23ft 4in)
Armament:	Six 533mm (21in) torpedo tubes, Harpoon missiles
Powerplant:	Single screw, diesel/electric motors
Surface range:	19,000km (10,241nm) at 9 knots
Performance:	Surfaced: 11 knots Submerged: 20 knots

Hajen

Hajen was the first submarine built for the Swedish Navy. She was designed by naval engineer Carl Richson, who had been sent to the USA to study submarine development in 1900. *Hajen* was laid down at Stockholm in 1902. In 1916 she underwent a major rebuild, and her length was increased by 1.8m (6ft). She was withdrawn from service in 1922, and became a museum exhibit. Sweden, which for centuries had been a major power in Europe, had adopted a policy of neutrality in the 1860s, after which it made every effort to retain a strong defensive navy. With the odd exception, all its warships, including submarines, were Swedish-built. Sweden's shipbuilding industry had an important asset in that it had access to high-grade steel, thanks to the country's quality iron ore deposits.

Country:	Sweden
Launch date:	July 1904
Crew:	15
Displacement:	Surfaced: 108 tonnes (107 tons)
	Submerged: 130 tonnes (127 tons)
Dimensions:	19.8m x 3.6m x 3m (65ft x 11ft 10in x 9ft 10in)
Armament:	One 457mm (18in) torpedo tube
Powerplant:	Single screw, paraffin engine, electric motor
Surface range:	Not known
Performance:	Surfaced: 9.5 knots
	Submerged: 7 knots

Han

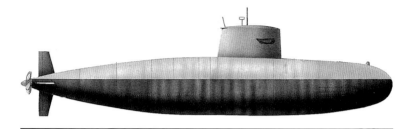

The Chinese Navy took a massive leap forward in the early 1970s with its Han-class nuclear-attack submarines (SSNs). The highly streamlined hull shape was based on the design of the USS *Albacore,* and was a radical departure from previous Chinese submarine designs. While the Russians cut many corners to get their first SSNs into service, China proceeded at a more leisurely pace, and although the Han class of four boats was fairly basic, compared with contemporary American and British vessels, it provided a solid basis for further development. Two of the five boat class have been decommissioned, with *Changzeng 1* preserved as a museum ship at the Chinese Navy Museum at Qingdao. From the Hans came the Xia class, which was China's first nuclear ballistic-missile submarine.

Country:	China
Launch date:	1972
Crew:	120
Displacement:	Surfaced: Not known Submerged: 5080 tonnes (5000 tons)
Dimensions:	90m x 8m x 8.2m (295ft 3in x 26ft 3in x 27ft)
Armament:	Six 533mm (21in) torpedo tubes
Powerplant:	Single screw, pressurized water nuclear reactor
Range:	Unlimited
Performance:	Surfaced: 20 knots Submerged: 28 knots

Harushio

Harushio was one of the Japanese Maritime Self-Defence Force's five Oshio-class submarines, completed in the mid-1960s. They were a large design in order to achieve improved seaworthiness and to allow the installation of more comprehensive sonar and electronic devices. They were the first Japanese submarines capable of deep diving. All five were built at Kobe, construction being shared between Kawasaki and Mitsubishi. The other boats in the class were *Oshio, Arashio, Michishio* and *Asashio*. The submarines were named after tides; *Oshio*, for example, means Flood Tide, while *Asashio* means Morning Tide. Because the lead boat, *Oshio*, had a different configuration from the others, with a larger bow and less sophisticated sonar, the boats are often cited as two separate classes.

Country:	Japan
Launch date:	25 February 1967
Crew:	80
Displacement:	Surfaced: (1676 tonnes) 1650 tons Submerged:
Dimensions:	88m x 8.2m x 4.9m (288ft 8in x 27ft x 16ft 2in)
Armament:	Eight 533mm (21in) torpedo tubes
Powerplant:	Two shafts, diesel/electic motors
Surface range:	16,677km (9000nm) at 10 knots
Performance:	Surfaced: 14 knots Submerged: 18 knots

Harushio

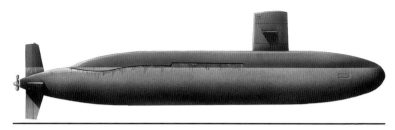

The seven submarines of the Harushio class were a natural progression from the previous Yushio class, with improved noise reduction and ZQR-1 towed array sonar. They were also equipped with the Hughes/Oki ZQQ-5B hull sonar. All were capable of firing the Sub-Harpoon anti-ship missile from their torpedo tubes. Beginning with *Harushio* in 1989, the boats were built at the rate of one a year to replace the vessels of the Uzushio class. The next five boats, in order of their launch date, were named *Natsushio, Hayashio, Arashio, Wakashio* and *Fuyushio*. A seventh boat, *Asashio*, was commissioned in 1997 but converted to a training submarine in 2000. She was the last boat to be decommissioned in 2017. The Harushio class were replaced in Japanese service by the larger Oyashio class equipped with a flank sonar array.

Country:	Japan
Launch date:	26 July 1989
Crew:	75
Displacement:	Surfaced: 2489 tonnes (2450 tons) Submerged:
Dimensions:	77m x 10m x 7.75m (252ft 7in x 32ft 10in x 25ft 4in)
Armament:	Six 533 mm (21in) torpedo tubes, Sub-Harpoon SSM
Powerplant:	Single shaft, diesel/electric motors
Surface range:	22,236km (12,000nm) at 10 knots
Performance:	Surfaced: 12 knots Submerged: 20 knots

Henri Poincaré

Built at Lorient, *Henri Poincaré* was one of 29 double-hulled ocean-going submarines of the Redoutable class laid down between 1925 and 1931. They were classified as ocean-going or first-class submarines, and were a considerable improvement over the preceding Requin class, but had their share of problems. *Prométhée,* was lost during trials on 8 July 1932, and another, *Phénix,* was lost in Indo-Chinese waters on 15 June 1939. *Henri Poincaré* was scuttled at Toulon in November 1942, along with her sister boats *Vengeur, Redoutable, Pascal, Achéron, L'Espoir* and *Fresnel,* but was salvaged by the Italians and returned to Genoa for overhaul. Designated *FR118*, she was sunk in September 1943 after being seized by German forces. France's Vichy-controlled submarines would have proved a valuable asset to the Allies, had they been released to the Free French.

Country:	France
Launch date:	10 April 1929
Crew:	61
Displacement:	Surfaced: 1595 tonnes (1570 tons)
	Submerged: 2117 tonnes (2084 tons)
Dimensions:	92.3m x 8.2m x 4.7m (302ft 10in x 27ft 15ft 5in)
Armament:	Nine 550mm (21.7in) and two 400mm (15.7in) torpedo tubes;
	one 82mm (3.2in) gun
Powerplant:	Twin screws, diesel/electric motors
Surface range:	18,530km (10,000nm) at 10 knots
Performance:	Surfaced: 17-20 knots
	Submerged: 10 knots

H L Hunley

HL Hunley was the first true submersible craft to be used successfully against an enemy. The main part of the hull was shaped from a cylindrical steam boiler, with the tapered ends added. Armament was a spar torpedo, an explosive charge on the end of a pole. The craft had a nine-man crew, eight to turn the handcranked propeller and one to steer. On 17 February 1864, commanded by Lt George Dixon, she slipped into Charleston Harbor and sank the newly commissioned Union corvette *Housatonic*, but was dragged down by the wave caused by the explosion of the torpedo. Years later, when the wreck was located on the sea bed, the skeletons of eight of the crew were discovered, still seated at their crankshaft. Named after her inventor, H L Hunley was one of a number of small submersibles built for the Confederate Navy.

Country:	Confederate States of America
Launch date:	1863
Crew:	9
Displacement:	Surfaced: 2 tonnes/2 tons approx Submerged: Not known
Dimensions:	12m x 1m x 1.2m (40ft x 3ft 6in x 4ft)
Armament:	One spar torpedo
Powerplant:	Single screw, hand-cranked
Surface range:	Not known
Performance:	Surfaced: 2.5 knots Submerged: Not known

Holland No 1

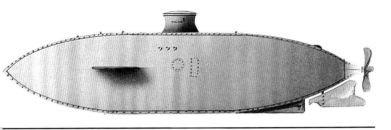

Although the Irish-American inventor John P. Holland had produced some experimental submarine designs, including the *Fenian Ram*, the *Holland No 1* was his first success. The diminutive craft was originally designed to be handcranked like previous submarines, but with the introduction of the newly-developed Brayton four-horsepower petrol engine, Holland was able to produce a more reliable vessel. *Holland No 1* was built at the Albany Iron Works and was completed in 1878. After successful trials the engine was removed and she was scuttled in 4.2m (14ft) of water on the Upper Passaic River. Years later she was raised, and she is now housed in the Paterson Museum, USA. Holland's far-sighted faith in the petrol engine proved premature, other submarine designs of this period being still dependent on steam for their motive power.

Country:	USA
Launch date:	1878
Crew:	Not known
Displacement:	surfaced: 2.2 tonnes/tons submerged:
Dimensions:	4.4m x 0.9m (14ft 6in x 3ft)
Armament:	None
Powerplant:	Single screw, petrol engine
Surface range:	Not known
Performance:	surfaced: Not known submerged: Not known

Holland VI

Holland VI was the first modern American submarine, and later became the prototype for British and Japanese submarines which combined petrol engine and battery power with hydroplanes. *Holland VI* entered service with the US Navy as the *Holland* in 1900. Her petrol engine developed 45hp and her electric motor 75hp when submerged. Diving depth was 22.8m (75ft). She served as a training boat until 1905, was re-numbered *SS1*, and scrapped in 1913. Although the American press praised the little Holland submarine and bestowed such lurid descriptions as 'Monster War Fish' on her, she was in fact a very primitive craft. She might not have been built at all, had Holland's political friends not persuaded the Navy to allocate funds to the venture. There was a very strong and influential Irish-American lobby in Congress at this time.

Country:	USA
Launch date:	May 1897
Crew:	7
Displacement:	Surfaced: 64 tonnes (63 tons) Submerged: 76 tonnes (74 tons)
Dimensions:	16.3m x 3.1m x 3.5m (53ft 3in x 10ft 3in x 11ft 6in)
Armament:	One 457mm (18in) torpedo tube; one pneumatic gun
Powerplant:	Single screw, petrol engine/electric motor
Range:	submerged: 74km (40nm) at 3 knots
Performance:	Surfaced: 8 knots Submerged: 5 knots

Hvalen

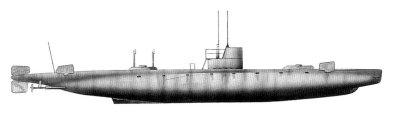

*H**valen** was the only large submarine to be constructed for Sweden by a foreign power. She was bought from the Italian firm of Fiat-San Giorgio so that the Swedes might have the opportunity to evaluate a leading European design, and she made headline news with her epic 7600km (4096nm) voyage from Italy to Sweden unescorted. *Hvalen* was removed from the effective list in 1919 and was sunk as a target in 1924. The wreck was later raised and scrapped. Because of her neutral status, Sweden has always been regarded as something of a backwater in naval affairs, but the fact remains that her shipyards have produced some of the world's best and most effective warships, including submarines. The Swedish government has tried to be self-sufficient in military and naval matters, recognising the dangers of relying on foreign partners.

Country:	Sweden
Launch date:	1909
Crew:	30
Displacement:	Surfaced: 189 tonnes (186 tons) Submerged: 233 tonnes (230 tons)
Dimensions:	42.4m x 4.3m x 2.1m (139ft x 14ft x 6ft 11in)
Armament:	Two 457mm (18in) torpedo tubes
Powerplant:	Single screw, petrol engines/electric motors
Surface range:	8338km (4500nm) at 10 knots
Performance:	Surfaced: 14.8 knots Submerged: 6.3 knots

I7

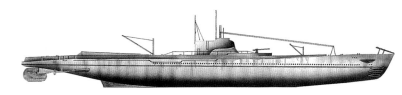

At the time of their construction, *I7* and her sister boat *I8* were the largest submarines built for the Japanese Navy. They were intended for the scouting role, and carried a reconnaissance seaplane. They could stay away from base for 60-day periods, cruising for over over 14,000nm at 16 knots, and could dive to a depth of 99m (325ft). Between them, the two boats sank seven Allied merchant ships totalling 42,574 (41,902 tons). *I7* was sunk by the American destroyer *Monaghan* on 22 June 1943. *I8* was modified to carry four Kaiten suicide submarines in place of her aircraft hangar. She was sunk by the destroyers USS *Morrison* and *Stockton* on 30 March 1945 while attempting to attack American ships involved in the Okinawa landings. Many Japanese warships were lost in suicide attacks during this campaign.

Country:	Japan
Launch date:	3 July 1935
Crew:	100
Displacement:	Surfaced: 2565 tonnes (2525 tons)
	Submerged: 3640 tonnes (3583 tons)
Dimensions:	109.3m x 9m x 5.2m (358ft 7in x 29ft 6in x 17ft)
Armament:	Six 533mm (21in) torpedo tubes, one 140mm (5.5in) gun
Powerplant:	Twin screws, diesel/electric motors
Surface range:	26,600km (14,337nm) at 16 knots
Performance:	Surfaced: 23 knots
	Submerged: 8 knots

I15

The I15-class submarines were highly specialized scouting boats, with streamlined hulls and conning towers. Compared with earlier classes of scouting submarines, the seaplane hangar was also streamlined, being a smooth, rounded fairing that extended forward as part of the conning tower. Although designed to mount four 25mm AA guns, they were completed with a single twin 25mm mounting. During World War II some vessels had their hangar and catapult removed and replaced by a second 140mm (5.5in) gun, being reclassed as attack submarines. Only one boat of the fairly large class, *I36*, survived the war, being surrendered at Kobe. *I15* (Cdr Ishikawa) was lost on 2 November 1942. As the war progressed and the Americans recaptured the Pacific islands the Japanese submarine fleet was penned into its home bases, its radius of action reduced.

Country:	Japan
Launch date:	1939
Crew:	100
Displacement:	Surfaced: 2625 tonnes (2584 tons) Submerged: 3713 tonnes (3654 tons)
Dimensions:	102.5m x 9.3m x 5.1m (336ft x 30ft 6in x 16ft 9in)
Armament:	Six 533mm (21in) torpedo tubes; one 140mm (5.5in) and two 25mm AA guns
Powerplant:	Two shafts, diesel/electric motors
Surface range:	45,189km (24,400nm) at 10 knots
Performance:	Surfaced: 23.5 knots Submerged: 8 knots

I21

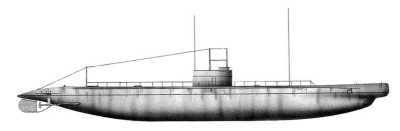

One of two vessels that were Japan's first ocean-going submarines, *I21* was built from Italian plans of the Fiat-Laurenti *F1* type. She was built at the Kawasaki yard in Kobe and completed in 1920. Her number was changed to *RO2* in 1924, and she was stricken in 1930. In the meantime, the *I21* number had been re-allocated to a new submarine, launched in March 1926; the design of this vessel was based on the German submarine *UB125*, which had been given to Japan after surrendering in 1918. The new *I21* was leader of a class of four boats, all of which went on to see service in the Pacific War. In 1939 they received new designations, *I21* becoming *I121* and so on. *I21/121* was scrapped; the others were lost in action. Very few Japanese submarines survived until the final surrender.

Country:	Japan
Launch date:	November 1919
Crew:	45
Displacement:	Surfaced: 728 tonnes (717 tons) Submerged: 1063 tonnes (1047 tons)
Dimensions:	65.6m x 6m x 4.2m (215ft 3in x 19ft 8in x 13ft 9in)
Armament:	Five 457mm (18in) torpedo tubes
Powerplant:	Two screws, diesel/electric motors
Surface range:	19,456km (10,500nm) at 8 knots
Performance:	Surfaced: 13 knots Submerged: 8 knots

I201

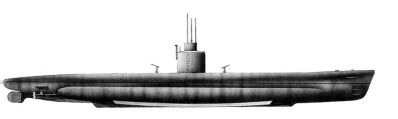

Ordered under an emergency 1943-44 construction programme, the I201 class of high-speed submarines was designed as a result of trials carried out with an experimental submarine just before World War II; they compared very favourably with the German Type XXI. They were highly streamlined, making extensive use of electric welding, and even the twin 25mm AA guns were on mounts that retracted into the hull. The lightweight MAN diesels contributed to a small displacement, which in turn gave rise to a fast submerged speed; the 5000hp electric motors could propel the boat under water at 19 knots for nearly an hour. None of the boats was able to carry out any operational patrols before the end of the war; all were scuttled by the US Navy or scrapped with the exception of *I204*, sunk in an air attack.

Country:	Japan
Launch date:	1944
Crew:	100
Displacement:	Surfaced: 1311 tonnes (1291 tons)
	Submerged: 1473 tonnes (1450 tons)
Dimensions:	79m x 5.8m x 5.4m (259ft 2in x 19ft x 17ft 9in)
Armament:	Four 533mm (21in) torpedo tubes
Powerplant:	Twin screws, diesel/electric motors
Surface range:	10,747km (5800nm) at 14 knots
Performance:	Surfaced: 15.7 knots
	Submerged: 19 knots

I351

The three submarines of the *I351* class were intended to act as supply bases for seaplanes and flying boats, for which role they were equipped to carry 396 tonnes (390 tons) of cargo, including 371 tonnes (365 tons) of petrol, 11 tons of fresh water and 60 250kg (550lb) bombs, or alternatively 30 bombs and 15 aircraft torpedoes. By the end of the war only the *I351* had been completed, and she was sunk by the American submarine USS *Bluefish* on 14 July 1945, after six months in service. A second boat, the *I352*, was sunk by air attack at Kure when she was 90 per cent complete, and a third, *I353*, was never laid down, having been cancelled in 1943. *I351* had a safe diving depth of 96m (315ft) and had an underwater range of 185km (100nm) at three knots. The role of these submarines was dictated by the loss of Japan's Pacific bases.

Country:	Japan
Launch date:	1944
Crew:	90
Displacement:	Surfaced: 3568 tonnes (3512 tons)
	Submerged: 4358 tonnes (4290 tons)
Dimensions:	110m x 10.2m x 6m (361ft x 33ft 6in x 19ft 8in)
Armament:	Four 533mm (12in) torpedo tubes
Powerplant:	Twin screws, diesel/electric motors
Surface range:	24,076km (13,000nm) at 14 knots
Performance:	Surfaced: 15.8 knots
	Submerged: 6.3 knots

I400

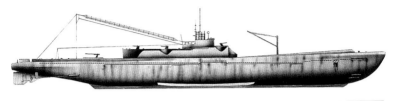

Prior to World War II, several navies tried to build an effective aircraft-carrying submarine. Only the Japanese managed to produce a series of workable vessels, the most notable being the STO class. Of the 19 planned vessels only two, the *I400* and *I401*, were completed for their intended role. A third, *I402*, was completed as a submersible tanker transport. *I400* was a huge vessel, with a large aircraft hangar offset to starboard, to hold three M6A1 Seiran floatplanes, plus components for a fourth. To launch the aircraft, *I400* would surface, then the machines would be warmed up in the hangar, rolled out, wings unfolded, and launched down a 26m (85ft) catapult rail. It was planned to attack the locks on the Panama Canal, but the mission was never flown. The I400 class was not rivalled in size until the emergence of the Ethan Allen class of SSBN.

Country:	Japan
Launch date:	1944
Crew:	100
Displacement:	Surfaced: 5316 tonnes (5233 tons) Submerged: 6665 tonnes (6560 tons
Dimensions:	122m x 12m x 7m (400ft 1in x 39ft 4in x 24ft)
Armament:	Eight 533mm (21in) torpedo tubes; one140mm (5.5in) gun
Powerplant:	Twin screws, diesel/electric motors
Surface range:	68,561km (37,000nm) at 14 knots
Performance:	Surfaced: 18.7 knots Submerged: 6.5 knots

India

*I**ndia** was designed for salvage and rescue operations. Her hull was built for high surface speeds, enabling her to be deployed rapidly to her rescue co-ordinates. Two rescue submarines are carried in semi-recessed deck wells aft, and personnel can enter the mother boat from these when she is submerged. The boat can also operate under ice. India-class submarines are believed to operate in support of Russian Spetsnaz special operations brigades when not being used in their primary role, carrying two IRM amphibious reconnaissance vehicles; these are capable of travelling along the sea bed on tracks as well as operating in thenormal swimming mode. Two Indias were built, and deployed with the Northern and Pacific Fleets, going to the aid of Russian nuclear submarines involved in accidents. Both were decommissioned in the mid-1990s.

Country:	Russia
Launch date:	1979
Crew:	70 (plus accommodation for 120 others)
Displacement:	Surfaced: 3251 tonnes (3200 tons)
	Submerged: 4064 tonnes (4000 tons)
Dimensions:	106m x 10m (347ft 9in x 32ft 10in)
Armament:	Four 533mm (21in) torpedo tubes
Powerplant:	Twin screws, diesel/electric motors
Surface range:	Not known
Performance:	Surfaced: 15 knots
	Submerged: 10 knots

Intelligent Whale

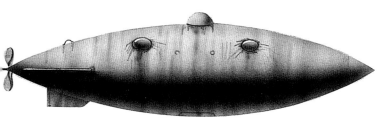

Intelligent Whale was the first submarine constructed for the Union Navy, and was built in response to Confederate vessels of the same type; for example the David (so named because it was seen as a giant-killer) which was steam-propelled and was really more of a semi-submersible torpedo boat than a true submarine. Thirteen men were carried in the *Whale,* six of them propelling the cylindrical cigar-shaped hull by hand, and the remainder intended to leave the vessel by a trap door in the floor in order to secure mines to the hulls of enemy vessels. After several tests, the project was finally abandoned in 1872 and *Intelligent Whale* was put on display at the Washington Navy Yard. In general, Confederate attempts to produce workable submersibles were more effective than those of their Federal opponents.

Country:	USA
Launch date:	1862
Crew:	13
Displacement:	Surfaced: Not known Submerged: Not known
Dimensions:	9.4m x 2.6m x 2.6m (31ft x 8ft 6in x 8ft 6in)
Armament:	Mines
Powerplant:	Single screw, hand-cranked
Surface range:	Not known
Performance:	Surfaced: 4 knots Submerged: 4 knots

Isaac Peral

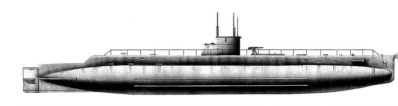

Isaac Peral was Spain's first major submarine. She was built by the Fore River Company in the United States and modelled on the Holland design. She attained 15.36 knots on the surface during trials. Surface range was 5386km (2835 statute miles and 2903 nautical miles) at 11 knots; range submerged was 130km (70nm) at full power from her 480hp electric motors. She was renumbered *O1* in 1930, later being reduced to a hulk and numbered *AO*. Her single 76mm (3in) gun was fixed to a collapsible mount, and was not a permanent feature. Spain did not maintain a substantial submarine force; after General Franco's victory in the Civil War, boats were obtained from Italy. The Civil War created an enormous drain on Spain's resources, one of the reasons why Franco chose to remain neutral in World War II.

Country:	Spain
Launch date:	July 1916
Crew:	35
Displacement:	Surfaced: 499 tonnes (491 tons)
	Submerged: 762 tonnes (750 tons)
Dimensions:	60m x 5.8m x 3.4m (196ft 10in x 19ft x 11ft 2in)
Armament:	Four 457mm (18in) torpedo tubes, one 76mm (3in) gun
Powerplant:	Twin screws, diesel/electric motors
Surface range:	5386km (2903nm) at 11 knots
Performance:	Surfaced: 15 knots
	Submerged: 8 knots

J1

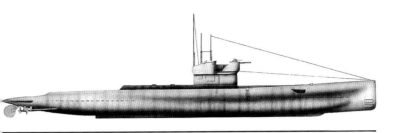

J1 was built in response to a perceived threat from German submarines then entering service and reputed to be capable of 22 knots. As first completed, *J1*'s large forward free-flooding tank brought the bows down in the water, causing loss of speed when surfaced. Later the bows were raised, curing this pitching tendency, and the submarine was then able to maintain 17 knots surfaced in heavy seas. Range at 12.5 knots surfaced was 9500km (5120nm). Later, a 102mm (4in) gun was positioned high up at the front of the conning tower in place of the 76mm (3in) weapon. On 5 November 1916, *J1* torpedoed and damaged the German battleships *Grosser-Kurfürst* and *Kronprinz*. *J1* was handed over to Australia in 1919, and was broken up in 1924. Only seven J-class boats were built, one of which was lost accidentally.

Country:	Britain
Launch date:	November 1915
Crew:	44
Displacement:	Surfaced: 1223 tonnes (1204 tons)
	Submerged: 1849 tonnes (1820 tons)
Dimensions:	84m x 7m x 4.3m (275ft 7in x 23ft x 14ft)
Armament:	Six 457mm (18in) torpedo tubes; one 76mm (3in) gun
Powerplant:	Triple screws, diesel/electric motors
Surface range:	9500km (5120nm) at 12.5 knots
Performance:	Surfaced: 17 knots
	Submerged: 9.5 knots

K4

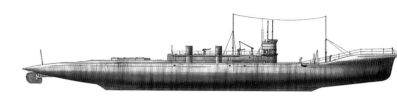

In 1915 the British Admiralty decided to design a class of exceptionally fast ocean-going submarines that could keep up with the battlefleet. As diesel engines of the period could not develop adequate power to sustain a surface speed of 24 knots, steam turbines were used instead, with electric motors for underwater operation. The turbine machinery took up nearly 40 per cent of a K boat's length, and had to be shut down when she was submerged, with large lids covering the funnel uptakes. The boats were a disaster, no fewer than five of the 17 built prior to 1919 being lost in accidents. It was hardly surprising that morale in the Submarine Flotillas to which the K boats were assigned was not at its highest level. In general, the K boats were relegated to anti-submarine patrols. *K4* was accidentally rammed and sunk by *K6* in February 1918.

Country:	Britain
Launch date:	15 July 1916
Crew:	50-60
Displacement:	Surfaced: 2174 tonnes (2140 tons)
	Submerged: 2814 tonnes (2770 tons)
Dimensions:	100.6m x 8.1m x 5.2m (330ft x 26ft 7in x 17ft)
Armament:	Ten 533mm (21in) torpedo tubes; three 102mm (4in) guns
Powerplant:	Twin screws, steam turbines/electric motors
Surface range:	5556km (3000nm) at 13.5 knots
Performance:	Surfaced: 23 knots
	Submerged: 9 knots

K26

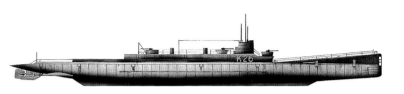

K*26* was the only one of the notorious K boats to be built after 1919, being completed in 1919 and scrapped in 1931. One K-boat incident might have cost the future King George VI his life. He was a passenger on board *K3* when the boat was being put through her paces. The submarine's commander prepared to dive but, instead of gently nosing under the surface, the boat went down at a steep angle, hit the sea bead and stuck fast, with her bows buried deep in mud. The water was about 46m (150ft) deep, so a sizeable portion of the boat was still protruding above the surface, her propellers still turning. Luckily she was released without damage after about 20 minutes of frantic effort. Other K-boat crews were not so lucky. None of the boats was destroyed in action; all five losses were caused by accidents.

Country:	Britain
Launch date:	August 1919
Crew:	50-60
Displacement:	Surfaced: 2174 tonnes (2140 tons) Submerged: 2814 tonnes (2770 tons)
Dimensions:	100.6m x 8.1m x 5.2m (330ft x 26ft 7in x 17ft)
Armament:	Ten 533mm (21in) torpedo tubes; three 102mm (4in) guns
Powerplant:	Twin screws, steam turbines/electric motors
Surface range:	5556km (3000nm) at 13.5 knots
Performance:	Surfaced: 23 knots Submerged: 9 knots

Kilo class

Built at Komsomolsk in the Russian Far East, the first medium-range Kilo-class submarine was launched early in 1980. By 1982 construction had also started at the Gorki shipyard, while export production began in 1985 at Sudomekh. In August 1985 the first operational Kilo deployed to the vast Vietnamese naval base at Cam Ranh Bay for weapon systems trials under tropical conditions, and in the following year the first sighting of a Kilo in the Indian Ocean was made by a warship of the Royal Australian Navy. The Kilo class had a more advanced hull form more typical of western 'teardrop' submarine hulls. Improved Kilo IIs are still being built for the Russian Navy, and Kilos have been exported to Algeria, China, India, Iran, Myanmar, Poland, Romania and Vietnam. One Kilo was heavily damaged by a Storm Shadow cruise missile while in Sevastapol drydock in 2023.

Country:	Russia
Launch date:	Early 1980 (first unit)
Crew:	45-50
Displacement:	Surfaced: 2494 tonnes (2455 tons) Submerged: 3193 tonnes (3143 tons)
Dimensions:	69m x 9m x 7m (226ft 5in x 29ft 6in x 23ft)
Armament:	Six 533mm (21in) torpedo tubes, Kalibr/Club SLCM, mines
Powerplant:	Single shaft, three diesels, three electric motors
Surface range:	11,112km (6000nm) at 7 knots
Performance:	Surfaced: 15 knots Submerged: 24 knots

L3

L3 was one of a large class of Russian submarines. On the night of 16 April 1943, commanded by Capt 3rd Class Konovalov, she intercepted a German convoy of eight ships evacuating refugees from the Hela peninsula in the Baltic to the west, and sank the large steamship *Goya*. Of the 6385 persons on board, only 165 were rescued. It was the climax of a long and successful operational career that began with minelaying operations in the Baltic in June 1941, days after the German invasion of Russia. Russian submarine operations in the Baltic were a considerable threat to German supply and reinforcement traffic. Minelaying continued to be *L3*'s principal occupation, and she did not register her first success until August 1942, when she sank the 5580-tonne (5492-ton) steamer *C.F. Liljevalch*. *L3* served for several years after the war, and was scrapped in 1959.

Country:	Russia
Launch date:	July 1931
Crew:	50
Displacement:	Surfaced: 1219 tonnes (1200 tons) Submerged: 1574 tonnes (1550 tons)
Dimensions:	81m x 7.5m x 4.8m (265ft 9in x 24ft 7in x 15ft 9in)
Armament:	Six 533mm (21in) torpedo tubes; one 100mm (3.9in) gun
Powerplant:	Twin screws, diesel/electric motors
Surface range:	11,112km (6000nm) at 9 knots
Performance:	Surfaced: 15 knots Submerged: 9 knots

L3

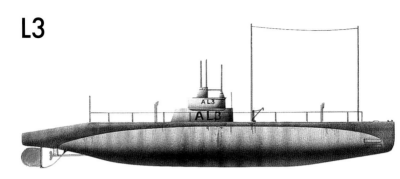

L 3 was the first American submarine to be fitted with a deck gun. This retracted vertically into a deckhouse until only a small portion of the barrel was left exposed, so reducing underwater drag. The United States ended World War I with about 120 submarines, although by this time she had lost her submarine design lead (established in the early years of the 20th century by John Holland) to the European naval powers. At this time, America's best submarines were roughly comparable with Britain's H and L classes. Very little progress in submarine design was made in the US during the inter-war years, and it took the threat of another war to act as the spur that would once again bring her to the forefront. Her ocean-going submarines soon achieved superiority in the Pacific.

Country:	USA
Launch date:	February 1915
Crew:	35
Displacement:	Surfaced: 457 tonnes (450 tons) Submerged: 556 tonnes (548 tons)
Dimensions:	51m x 5.3m x 4m (167ft 4in x 17ft 4in x 13ft 1in)
Armament:	Four 457mm (18in) torpedo tubes; one 76mm (3in) gun
Powerplant:	Twin screws, diesel/electric motors
Surface range:	6270km (3380nm) at 11 knots
Performance:	Surfaced: 14 knots Submerged: 8 knots

L10

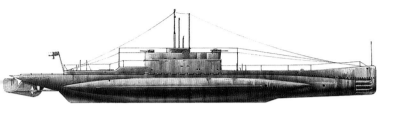

After the Battle of Jutland in May 1916, German surface forces rarely put to sea, and targets for Allied submarines were few and far between. Then, in April 1918, the German High Seas Fleet once more ventured out in strength to attack convoys plying between Britain and Scandinavia, and these activities continued, albeit on a reduced scale, until almost the end of the war, with destroyers the main participants. During one of these sorties, on 3 October 1918, the British submarine *L10* intercepted the German destroyer *S33* and sank her – only to be sunk herself by other enemy warships. During these last weeks of the war the German destroyers were very active, and British submarines often became the hunted rather than the hunters. It was an indication of what determined surface forces could do in the anti-submarine war.

Country:	Britain
Launch date:	24 January 1918
Crew:	36
Displacement:	Surfaced: 904 tonnes (890 tons) Submerged: 1097 tonnes (1080 tons)
Dimensions:	72.7m x 7.2m x 3.4m (238ft 6in x 23ft 8in x 11ft 2in)
Armament:	Four 533mm (21in) torpedo tubes; one 102mm (4in) gun
Powerplant:	Twin screws, diesel/electric motors
Surface range:	7038km (3800nm) at 10 knots
Performance:	Surfaced: 17.5 knots Submerged: 10.5 knots

L23

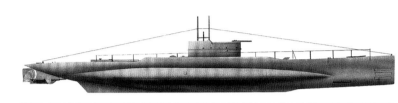

L*23* was one of the last surviving units of the large L class of submarines, 17 of
which were built after the end of World War I. One of the L-class boats, *L12*,
torpedoed and sank the German submarine *UB90* on 16 October 1918 while the
enemy boat was recharging her batteries on the surface of the North Sea at night.
The second boat of the class, *L2*, was subjected to a fierce gunfire and depth
charge attack by American warships escorting a convoy in February 1918, one
shell scoring a direct hit on the pressure hull just behind the conning tower as
the boat re-surfaced. Fortunately, the Americans realised their mistake in time
to avert a tragedy. Three boats, *L23*, *L26* and *L27*, served on training duties in
World War II; *L23* foundered under tow off Nova Scotia en route to the breaker's
yard in May 1946.

Country:	Britain
Launch date:	1 July 1919
Crew:	36
Displacement:	Surfaced: 904 tonnes (890 tons) Submerged: 1097 tonnes (1080 tons)
Dimensions:	72.7m x 7.2m x 3.4m (238ft 6in x 23ft 8in x 11ft 2in)
Armament:	Four 533mm (21in) torpedo tubes; one 102mm (4in) gun
Powerplant:	Twin screws, diesel/electric motors
Surface range:	8338km (4500nm)
Performance:	Surfaced: 17.5 knots Submerged: 10.5 knots

Los Angeles

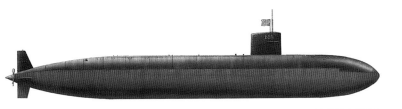

The lead ship of this class, *Los Angeles*, was commissioned on 13 November 1976. She was followed by 61 more, the last of which, *Cheyenne,* was commissioned on 13 September 1996. They are nuclear attack submarines (SSN), fulfilling a variety of roles: land attack with onboard Tomahawk cruise missiles, anti-ship with the Harpoon SSM and anti-submarine torpedoes, first fired by the USS *Norfolk* on 23 July 1988. Nine of the class were involved in the Gulf War of 1991, two firing Tomahawk missiles at targets in Iraq from stations in the eastern Mediterranean. From SN719 (USS *Providence*) onwards all were equipped with the vertical launch system, 12 launch tubes external to the pressure hull. The last 23 were built to an improved 688i standard. As of 2024, 24 were still active with the US Navy, with the last examples expected to serve until at least the late 2020s.

Country:	USA
Launch date:	6 April 1974
Crew:	133
Displacement:	Surfaced: 6180 tonnes (6082 tons) Submerged: 7038 tonnes (6927 tons)
Dimensions:	110.3m x 10.1m x 9.9m (362ft x 33ft x 32ft 3in)
Armament:	Four 533mm (21in) torpedo tubes; Tomahawk Land Attack Cruise Missiles, Harpoon SSM
Powerplant:	Single shaft, nuclear PWR, turbines
Range:	Unlimited
Performance:	Surfaced: 20 knots Submerged: 32 knots

Luigi Settembrini

Luigi Settembrini was a fast, short-range, partial double-hull boat with excellent manoeuvrability. Until 1940 she served in the Red Sea. Fron 1940 to 1943 she alternated combat patrols (on which she proved completely ineffective) and supply runs to North Africa with periods of service at the Italian Navy's Submarine School. After Italy joined the Allies, the boat was used for training until she was accidentally rammed and sunk by the US escort destroyer *Framet*. *Settembrini's* sister boat, *Ruggiero Settimo,* followed much the same operational career; she was launched in March 1931, completed in April 1932, and was eventually stricken from the Navy list on 23 March 1947. Very few of the wartime Italian submarines were judged effective enough to continue in a combat role in the post-war years.

Country:	Italy
Launch date:	28 September 1930
Crew:	56
Displacement:	surfaced: 968 tonnes (953 tons) submerged: 1171 tonnes (1153 tons)
Dimensions:	69m x 6.6m x 4.4m (226ft 8in x 21ft 8in x 14ft 5in)
Armament:	Eight 533mm (21in) torpedo tubes; one 102mm (4in) gun
Powerplant:	Twin screws, diesel/electric motors
Surface range:	16,668km (9000nm) at 8 knots
Performance:	surfaced: 17 knots submerged: 7.5 knots

M1

In 1917 the British Admiralty suspended construction work on four K boats and revised their plans, to turn them into submarine monitors, known as the M class, by mounting a single 305mm (12in) gun in the front part of an extended conning tower. The gun could be fired from periscope depth within 30 seconds of a target being sighted, or in 20 seconds if the submarine was surfaced. The snag was that the gun could not be reloaded under water, so the submarine had to surface after each round was fired – earning the M class the nickname 'Dip Chicks'. The M boats were intended to be the equivalent of Germany's 'cruiser submarines'. They were never used operationally; three were completed and two lost in accidents, the *M1* herself being lost in November 1925 when she collided with the freighter *Vidar*.

Country:	Britain
Launch date:	9 September 1917
Crew:	60-70
Displacement:	Surfaced: 1619 tonnes (1594 tons) Submerged: 1977 tonnes (1946 tons)
Dimensions:	90m x 7.5m x 4.9m (295ft 7in x 24ft 7in x 16ft)
Armament:	Four 533mm (21in) torpedo tubes; one 305mm (12in) gun
Powerplant:	Twin screws, diesel/electric motors
Surface range:	7112km (3840nm) at 10 knots
Performance:	Surfaced: 15 knots Submerged: 9 knots

Marlin

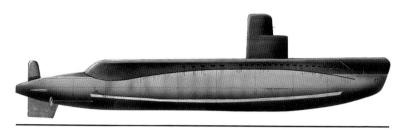

T he target submarines (SSTs) *Mackerel* and *Marlin* were authorized in the Fiscal Year 1951 and 1952 shipbuilding programmes, respectively. They were the smallest American submarines built since the C class of 1909, and were intended specifically for anti-submarine training. During 1966-67 the *Mackerel* evaluated equipment for the deep submergence vehicle NR-1, including keel-mounted wheels for rolling over the ocean floor, thrusters, external television cameras, a manipulator arm, and experimental sonar. The *Mackerel* 'bottomed' 225 times during the nine-month evaluation. *Mackerel* and *Marlin* were originally designated *T-1* and *T-2*, being named in 1956. *Mackerel* was built by the Electric Boat Company and *Marlin* by the Portsmouth Naval Yard; both were stricken in January 1973.

Country:	USA
Launch date:	17 July 1953
Crew:	18
Displacement:	Surfaced: 308 tonnes (303 tons) Submerged: 353 tonnes (347 tons)
Dimensions:	40m x 4.1m x 3.7m (131ft 2in x 13ft 6in x 12ft 2in)
Armament:	One 533mm (21in) torpedo tube
Powerplant:	Single shaft, diesel/electric motors
Surface range:	3706km (2000nm) at 8 knots
Performance:	Surfaced: 8 knots Submerged: 9.5 knots

Marsopa

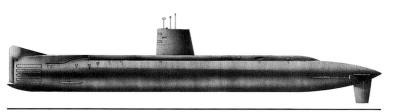

Marsopa is one of four Daphné-class boats of French design manufactured under licence in Spain, the others being *Delfin, Tonina* and *Narval.* All the Spanish boats underwent updates and modifications similar to those applied to the French vessels. The Daphné class, dogged by misfortune in its early operational days, went on to be an export success; in addition to the 11 French and four Spanish units, Portugal received the *Albacore, Barracuda, Cachalote* and *Delfin,* while Pakistan took delivery of the *Hangor, Shushuk* and *Mangro* (and also, later, Portugal's *Cachalote,* which was renamed *Ghazi*). In 1971 the *Hangor* made the first submarine attack since World War II, sinking the Indian frigate *Khukri* during the Indo-Pakistan war. South Africa also received three boats of this type. *Cachalote* was the last operational Daphné, stricken only in 2010.

Country:	Spain
Launch date:	15 March 1974
Crew:	45
Displacement:	Surfaced: 884 tonnes (870 tons) Submerged: 1062 tonnes (1045 tons)
Dimensions:	58m x 7m x 4.6m (189ft 8in x 22ft 4in x 15ft)
Armament:	Twelve 552mm (21.7in) torpedo tubes
Powerplant:	Two diesels, two electric motors
Surface range:	8338km (4300nm) at 5 knots
Performance:	Surfaced: 13.5 knots Submerged: 16 knots

N1

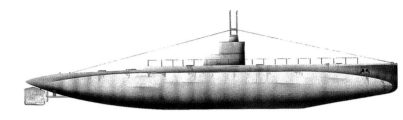

When the USA entered World War I in 1917 there were around 50 submarines in service with the US Navy. These ranged from the small A and B boats in the Philippines to the more advanced boats of the L class. The seven units of the N class were slightly smaller than the previous L class, and had reduced engine power in order to achieve greater engine reliability. This led to the adoption of more moderate power in the subsequent O, R and S classes, the last of which were launched in 1922. *N1* and her class were the first US submarines to have metal bridges, and the last until 1946 to be designed without deck guns. *N1*, re-numbered *SS53* in 1920, was broken up in 1931. The US Navy's submarines were used mainly in the coastal defence role in World War I, as a result of their restricted endurance.

Country:	USA
Launch date:	December 1916
Crew:	35
Displacement:	Surfaced: 353 tonnes (348 tons) Submerged: 420 tonnes (414 tons)
Dimensions:	45m x 4.8m x 3.8m (147ft 4in x 15ft 9in x 12ft 6in
Armament:	Four 457mm (18in) torpedo tubes
Powerplant:	Twin screws, diesel/electric motors
Range:	(submerged) 6485km (3500nm) at 5 knots
Performance:	Surfaced: 13 knots Submerged: 11 knots

Näcken

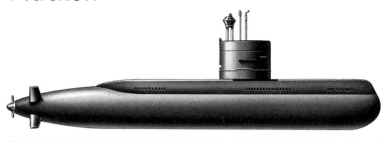

Built under a March 1973 contract by the Kockums and Karlskrona Navy Yard, the A14-type attack submarine *Näcken* and her two sisters, *Neptun* and *Najad,* were fitted with Kollmorgen periscopes. They had two decks and a 'teardrop' hull design. Their Saab NEDPS system used two Censor 392 computers to give engine and tactical information. In 1987–88 *Näcken* was fitted with two United Stirling Type V4-275 closed-cycle engines, increasing her overall length by 8m (26ft 3in). The powerplant could operate without the need for oxygen, so *Näcken* could remain submerged for up to 14 days. The three boats were specifically designed to counter incursions into Swedish territorial waters by Russian Whiskey-class diesel/electric boats, which continued at regular intervals during the Cold War era. All were decommissioned by 2016.

Country:	Sweden
Launch date:	17 April 1978
Crew:	19
Displacement:	Surfaced: 996 tonnes (980 tons) Submerged: 1168 tonnes (1150 tons)
Dimensions:	44m x 5.7m x 5.5m (144ft 4in x 18ft 8in x 18ft)
Armament:	Six 533mm (21in) and two 400mm (15.7in) torpedo tubes
Powerplant:	Single shaft, diesel/electric motors
Surface range:	3335km (1800nm) at 10 knots
Performance:	Surfaced: 20 knots Submerged: 25 knots

Narwhal

The USS *Narwhal* (SSN671) was constructed in 1966–67 to evaluate the natural-circulation SSG nuclear reactor plant. This uses natural convection rather than several circulator pumps, with their associated electrical and control equipment, for heat transfer operations via the reactor coolant to the steam generators, effectively reducing at low speeds one of the major sources of self-generated machinery noise within ordinary nuclear reactor-powered submarines. In all other respects the boat was similar to the Sturgeon-class SSNs. The other SSN built for powerplant research purposes was the USS *Glenard P. Lipscomb*, which evaluated a turbine-electric drive propulsion unit. Thanks to research of this kind, the American SSNs were much quieter than their Russian counterparts. Both submarines were fully operational units in the Atlantic until the late 1980s.

Country:	USA
Launch date:	9 September 1967
Crew:	141
Displacement:	Surfaced: 4521 tonnes (4450 tons) Submerged: 5436 tonnes (5350 tons)
Dimensions:	95.9m x 11.6m x 7.9m (314ft 8in x 38ft x 25ft 11in)
Armament:	Four 533mm (21in) torpedo tubes; SUBROC and Sub Harpoon missiles
Powerplant:	Single shaft, nuclear PWR, turbines
Range:	Unlimited
Performance:	Surfaced: 18 knots Submerged: 26 knots

Nautilus

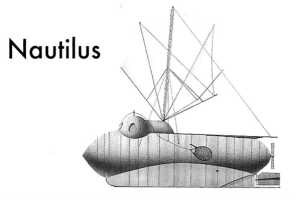

Nautilus was designed by Robert Fulton. Having no success with his project in America, Fulton went to France in 1797, where his plans for *Nautilus* were accepted. She was to become the first submarine to be built to a government contract. Her hull comprised an iron framework covered with copper sheets, and buoyancy was controlled by hand pumps. *Nautilus* was propelled by a sail when on the surface, and by hand-cranked propeller when submerged. During trials in Le Havre harbour, *Nautilus* remained under water at a depth of 7.6m (25ft) for one hour. A detachable explosive charge was secured just above the conning tower, ready to be fastened to the hull of an enemy vessel. After the French lost interest in the project Fulton took his design to Britain, but despite the fact that *Nautilus* was consistently successful during trials, she was not adopted.

Country:	USA/France
Launch date:	1800
Crew:	3
Displacement:	Surfaced: 19 tonnes /tons Submerged: Not known
Dimensions:	6.4m x 1.2m (21ft x 3ft 7in)
Armament:	Single detachable explosive charge
Powerplant:	Single screw, hand-cranked
Surface range:	Not known
Performance:	Surfaced: Not known Submerged: Not known

Nautilus

Nautilus was built in response to a request from the British Admiralty for a 1016-tonne (1000-ton) submarine with a surface speed of 20 knots. The idea was that such a craft could accompany units of the fleet and, if necessary, act in a protective role. However, calculations showed that the best speed that could be achieved was 17 knots, and that on a displacement of 1290 tonnes (1270 tons). Vickers, the designers, approached Fiat, but the latter firm could not guarantee the required 1850hp from their new 12-cylinder diesel engines. Nevertheless, *Nautilus* was laid down in 1913 and the big diesel engines were installed. She never entered operational service, but was stationed at Portsmouth as a depot ship. This apparent failure conceals the fact that *Nautilus* was a giant step forward, not only in size but in power. She was scrapped in 1922.

Country:	Britain
Launch date:	December 1914
Crew:	35
Displacement:	Surfaced: 1464 tonnes (1441 tons) Submerged:
Dimensions:	78.8m x 7.9m x 5.4m (258ft 6in x 26ft x 17ft 9in)
Armament:	Eight 457mm (18in) torpedo tubes
Powerplant:	Twin screws, diesel/electric motors
Surface range:	9816km (5300nm) at 11 knots
Performance:	Surfaced: 17 knots Submerged: 10 knots

Nautilus

The USS *Nautilus* (SS168, formerly *V6*) was one of three V-class boats designed as long-endurance ocean-going vessels with a heavy armament, the others being *Argonaut* and *Narwhal*. In 1940 she was re-fitted to carry 5104 litres (19,320 gallons) of aviation fuel for long-range seaplanes. In June 1942 she was one of a force of submarines patrolling north-west of Midway Island to counter an anticipated Japanese invasion force, and in August, together with Argonaut, she landed US raiders on Makin, in the Gilbert Islands. In October 1942 she sank two freighters off the east coast of Japan, and in May 1943. Together with *Narwhal*, she acted as a marker submarine in Operation Landcrab, the reconquest of Attu in the North Pacific. In March 1944 she sank another large freighter off the Mandate Islands. She was scrapped in 1945.

Country:	USA
Launch date:	15 March 1930
Crew:	90
Displacement:	Surfaced: 2773 tonnes (2730 tons) Submerged: 3962 tonnes (3900 tons)
Dimensions:	113m x 10m x 4.8m (370ft x 33ft 3in x 15ft 9in)
Armament:	Six 533mm torpedo tubes, two 152mm (6in) guns
Powerplant:	Twin screws, diesel/electric motors
Surface range:	33,336km (18,000nm) at 10 knots
Performance:	Surfaced: 17 knots Submerged: 8 knots

Nautilus

Nautilus was the world's first nuclear-powered submarine. Apart from her revolutionary propulsion system, she was a conventional design. Early trials established new records, including nearly 2250km (1213nm) submerged in 90 hours at 20 knots, at that time the longest period spent underwater by an American submarine, as well as being the fastest speed submerged. There were two prototype nuclear attack submarines; the other, USS *Seawolf,* was launched in July 1955, the last US submarine to feature a traditional conning tower, as distinct from the fin of later nuclear submarines. *Nautilus* was the more successful; *Seawolf* was designed around the S2G reactor, intended as a backup to the S2W, but it had many operational problems and was replaced by an S2W in 1959. *Nautilus* was preserved as a museum exhibit at Groton, Connecticut, in 1982.

Country:	USA
Launch date:	21 January 1954
Crew:	105
Displacement:	Surfaced: 4157 tonnes (4091 tons)
	Submerged: 4104 tonnes (4040 tons)
Dimensions:	97m x 8.4m x 6.6m (323ft 7in x 27ft 8in x 21ft 9in)
Armament:	Six 533mm (21in) torpedo tubes
Powerplant:	Twin screws, one S2W reactor, turbines
Range:	Unlimited
Performance:	Surfaced: 20 knots
	Submerged: 23 knots

Nazario Sauro

During the early 1970s it became apparent to the Supermarina (Italian Admiralty) that a new submarine design was required for defence against amphibious landings and for ASW and anti-shipping tasks in the local area. The result was the Sauro class. The first two units were the *Nazario Sauro* and *Carlo Fecia de Cossato,* which entered service in 1980 and 1979 respectively following a delay caused by major problems with their batteries. A further two units, the *Leonardo da Vinci* and *Guglielmo Marconi,* were then ordered and these were commissioned into service in 1981 and 1982. Two more boats *Salvatore Pelosi* and *Giuliano Prini,* followed in 1988 and 1989, with the final two, *Primo Longobardo* and *Gianfranco Gazzana Priaroggia,* commissioned in 1993 and 1995. The four oldest boats were decommissioned between 2002 and 2010.

Country:	Italy
Launch date:	9 October 1976
Crew:	45
Displacement:	Surfaced: 1479 tonnes (1456 tons)
	Submerged: 1657 tonnes (1631 tons)
Dimensions:	63.9m x 6.8m x 5.7m (209ft 7in x 22ft 4in x 18ft 8in)
Armament:	Six 533mm (21in) torpedo tubes
Powerplant:	Single shaft, diesel/electric motors
Surface range:	12,971km (7000nm) at 10 knots
Performance:	Surfaced: 11 knots
	Submerged: 20 knots

Nereide

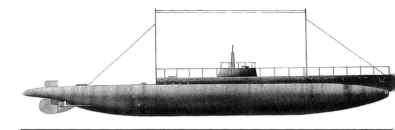

Nereide and her sister *Nautilus* were the first submarines designed by
Engineer Bernardi, later to become renowned as a submarine designer. Both
were laid down on 1 August 1911 and completed in 1913. *Nereide's* smooth, sleek
hull shape was similar to that of a torpedo boat, and her two torpedo tubes were
mounted in the bow. A third, deck-mounted torpedo tube was originally planned,
but never fitted. *Nereide* was sunk on 5 August 1915 near Pelagosa Island in the
Adriatic, by torpedoes from the Austrian submarine *U5*. Apart from a few loaned
by Germany, Austria's submarines were built in the country's own yards, and
were very compact boats. The majority of Italian submarines designed around
this period were efficient, capable boats, and were well suited to the waters of the
Adriatic, their principal operating area.

Country:	Italy
Launch date:	July 1913
Crew:	35
Displacement:	Surfaced: 228 tonnes (225 tons)
	Submerged: 325 tonnes (320 tons)
Dimensions:	40m x 4.3m x 2.8m (134ft 2in x 14ft 1in x 9ft 2in)
Armament:	Two 450mm (17in) torpedo tubes
Powerplant:	Twin screws, diesel/electric motors
Surface range:	7412km (4000nm) at 10 knots
Performance:	Surfaced: 13.2 knots
	Submerged: 8 knots

Nordenfelt 1

Nordenfelt was British-designed and built at Landskrona in Sweden. Laid down in 1882, she was one of the world's first steam-powered submarines. Her hull was almost circular in section, with frames spaced 0.6m (2ft) apart along its length. Diving depth was 15m (50ft). Most of her interior was taken up with the machinery, boilers and a steam accumulator, which had a heat exchanger at the bottom. Steam from the boiler was conveyed through the coils of the heater, giving up its latent heat to the water in the accumulator, which was then returned to the boiler via a feed pump. By this means, a large amount of superheated water could be stored in the pear-shaped tank. When this was released into the main boiler at a lower pressure, it turned into steam. The engine was fired in harbour, and took three days to heat up the reservoir fully.

Country:	Greece
Launch date:	1885
Crew:	Not known
Displacement:	Surfaced: 61 tonnes (60 tons) Submerged: Not known
Dimensions:	19.5m x 2.7m (64ft x 9ft)
Armament:	One 355mm (14in) gun; one 25.4mm (1in) gun (fitted later)
Powerplant:	Single screw, compound engine
Surface range:	Not known
Performance:	Surfaced: 9 knots Submerged: 4 knots

November class

Built as Russia's first nuclear submarine design from 1958 to 1963 at Severodvinsk, the November-class SSN was designed for the anti-ship rather than the anti-submarine role. They carried a full load of 24 nuclear torpedoes. They were provided with a targeting radar for use with the strategic-attack torpedo, presmably to confirm its position off an enemy coast. Thus armed, the task of these boats was to attack carrier battle groups. They were very noisy underwater and were prone to reactor leaks, which did not endear them to their crews. In April 1970 a November-class boat was lost south-west of the United Kingdom after an internal fire, the surviving crew being taken off before the boat sank, and there were numerous other incidents involving the boats during their operational career. All 14 boats were retired in the 1980s.

Country:	Russia
Launch date:	1958
Crew:	86
Displacement:	Surfaced: 4267 tonnes (4200 tons) Submerged: 5080 tonnes (5000 tons)
Dimensions:	109.7m x 9.1m x 6.7m (359ft 11in x 29ft 10in x 22ft)
Armament:	Eight 533mm (21in) and two 406mm (16in) torpedo tubes
Powerplant:	Twin shafts, one nuclear PWR, two turbines
Range:	Unlimited
Performance:	Surfaced: 20 knots Submerged: 30 knots

Nymphe

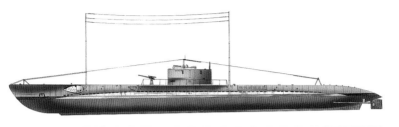

On the outbreak of World War II, the group of medium-range submarines to which *Nymphe* belonged was the largest class of such vessels in the French Navy, and they operated intensively until the French collapse in June 1940. French submarines were equally useful on the surface or submerged. *Nymphe* was one of a batch of four boats built by Ateliers Loire-Simonot; they were laid down in 1923 and completed in 1927. In spite of having a complex torpedo layout, involving a double revolving mounting, *Nymphe* and her consorts were successful ships. Three of the batch were scuttled at Toulon on 27 November 1942, shortly before the harbour was occupied by troops of the II SS Armoured Corps, this action following the Allied landings in North Africa earlier in the month. *Nymphe* herself was scrapped in 1938.

Country:	France
Launch date:	1 April 1926
Crew:	41
Displacement:	Surfaced: 619 tonnes (609 tons) Submerged: 769 tonnes (757 tons)
Dimensions:	64m x 5.2m x 4.3m (210ft x 17ft x 14ft)
Armament:	Seven 551mm (21.7in) torpedo tubes; one 76mm (3in) gun
Powerplant:	Twin screw diesel/electric motors
Surface range:	6485km (3500nm) at 7.5 knots
Performance:	Surfaced: 13.5 knots Submerged: 7.5 knots

O class

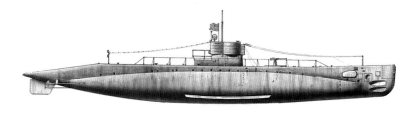

The 16 O-class submarines were all built for the US Navy in 1917–18. Only one, *O7*, saw active service in World War I, carrying out war patrols off the eastern seaboard of the United States from July 1918. Of the others, *O11*, *O13*, *O14*, *O15* and *O16* were broken up in 1930; *O2*, *O3*, *O4*, *O6*, *O7*, *O8* and *O10* were broken up in 1946: *O5* sank on 28 October 1923; *O9* went missing on 20 June 1941; and *O1* was stricken in 1938. One of the boats, *O12*, was sold to Norway in 1930, renamed *Nautilus* and used in an abortive attempt to reach the North Pole. She was broken up in 1931. During their service lives most of the O-class boats were used for training. They had a designed diving depth of 61m (200ft). Because of their restricted endurance, American submarines were generally confined to coastal patrol duties in World War I.

Country:	USA
Launch date:	9 July 1918 (O1)
Crew:	29
Displacement:	Surfaced: 529 tonnes (521 tons) Submerged: 639 tonnes (629 tons)
Dimensions:	52.5m x 5.5m x 4.4m (172ft 3in x 18ft 1in x 14ft 5in)
Armament:	Four 457mm (18in) torpedo tubes; one 76mm (3in) gun
Powerplant:	Two shafts, diesel/electric motors
Surface range:	10,191km (5500nm) at 11.5 knots
Performance:	Surfaced: 14 knots Submerged: 10.5 knots

Oberon

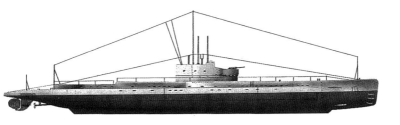

HM submarine *Oberon* was an ocean-going saddle-tank boat developed from the L-type minelaying submarines of World War I. She was laid down in 1924 and completed in 1927. Originally designated *O1*, the boat was of an advanced design and had a very respectable radius of action, which made the O class ideal for service in Far Eastern waters. Of the first three boats, however, only *Oxley* got that far, serving with the Royal Australian Navy from 1927 to 1931 before returning to home waters, where she was lost on 10 September 1939 after being rammed in error by HM submarine *Triton*. A second group of six O-class boats was built in 1928–29, four of these serving in the East Indies before being transferred to the Mediterranean in 1940. Only two of this batch survived the war. *Oberon* was scrapped at Rosyth in August 1945.

Country:	Britain
Launch date:	24 September 1926
Crew:	54
Displacement:	Surfaced: 1513 tonnes (1490 tons) Submerged: 1922 tonnes (1892 tons)
Dimensions:	83.4m x 8.3m x 4.6m (273ft 8in x 27ft 3in x 15ft)
Armament:	Eight 533mm (21in) torpedo tubes
Powerplant:	Twin screws, diesel/electric motors
Surface range:	9500km (5633nm) at 10 knots
Performance:	Surfaced: 13.7 knots Submerged: 7.5 knots

Oberon

Built between 1959 and 1967 as a follow-on to the Porpoise class, the Oberon class was outwardly identical to its predecessor, although there were internal differences. These included the soundproofing of all equipment for silent running and the use of a high-grade steel for the hull to allow a greater maximum diving depth of up to 340m (1115ft). A total of 13 units was commissioned into the Royal Navy. *Oberon* was modified with a deeper casing to house equipment for the initial training of personnel for the nuclear submarine fleet but was paid off for disposal in 1986, together with HMS *Orpheus*. One of the class, *Onyx*, served in the South Atlantic during the Falklands war on periscope beach recon-naissance operations and for landing special forces. During these operations she rammed a rock, causing one of her torpedoes to become stuck in its tube.

Country:	Britain
Launch date:	18 July 1959
Crew:	69
Displacement:	Surfaced: 2063 tonnes (2030 tons) Submerged: 2449 tonnes (2410 tons)
Dimensions:	90m x 8.1m x 5.5m (295ft 3in x 26ft 6in x 18ft)
Armament:	Eight 533mm (21in) torpedo tubes
Powerplant:	Two shafts, two diesel/electric motors
Surface range:	11,118km (6000nm) at 10 knots
Performance:	Surfaced: 12 knots Submerged: 17.5 knots

Odin

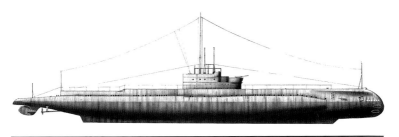

Odin was the lead boat of the second batch of O-class submarines built in the late 1920s. The O-class boats had a fairly heavy armament, but at the expense of handling quality. Together with *Olympus, Orpheus* and *Otus,* she served in the East Indies and then went to the Mediterranean in 1940, joining other O-class boats already there. On 14 June 1940, only four days after Italy's entry into the war and while operating out of Malta, she was sunk in the Gulf of Taranto by the Italian destroyer *Strale.* Her sister boat *Orpheus* was sunk off Tobruk by the destroyer *Turbine* two days later, *Oswald* was sunk by the destroyer *Vivaldi* south of Calabria on 1 August 1940, and *Olympus* was mined off Malta on 8 May 1942. The other two boats, *Osiris* and *Otus,* were scrapped at Durban in September 1946.

Country:	Britain
Launch date:	5 May 1928
Crew:	54
Displacement:	Surfaced: 1513 tonnes (1490 tons) Submerged: 1922 tonnes (1892 tons)
Dimensions:	83.4m x 8.3m x 4.6m (273ft 8in x 27ft 3in x 15ft)
Armament:	Eight 533mm (21in) torpedo tubes
Powerplant:	Twin screws, diesel/electric motors
Surface range:	9500km (5633nm) at 10 knots
Performance:	Surfaced: 17.5 knots Submerged: 8 knots

Ohio

The USS *Ohio* was the lead ship of a large class of nuclear-missile submarines (SSBN) intended to form the third arm of America's nuclear triad. *Ohio* was commissioned in November 1981. Boats of the class can remain submerged for up to 70 days. Four Ohio-class boats were converted to SSGNs from 2002–8. These were the *Ohio* (SSGN 726), *Michigan* (SSGN 727), *Florida* (SSGN 728), *Georgia* (SSGN 729). The remaining 14 boats are *Henry M. Jackson* (SSBN 730), *Alabama* (SSBN 731), *Alaska* (SSBN 732), *Nevada* (SSBN 733), *Tennessee* (SSBN 734), *Pennsylvania* (SSBN 735), *West Virgina* (SSBN 736), *Kentucky* (SSBN 737), *Maryland* (SSBN 738), *Nebraska* (SSBN 739), *Rhode Island* (SSBN 740), *Maine* (SSBN 741), *Wyoming* (SSBN 742) and *Louisiana* (SSBN 743). All USN SSBNs are under the control of USAF Strategic Air Command.

Country:	USA
Launch date:	7 April 1979
Crew:	155
Displacement:	Surfaced: 16,360 tonnes (16,764 tons) Submerged: 19,050 tonnes (18,750 tons)
Dimensions:	170.7m x 12.8m x 11m (560ft x 42ft x 36ft 5in)
Armament:	24 Trident C4 missiles, four 533mm (21in) torpedo tubes
Powerplant:	Single shaft, nuclear PWR
Range:	Unlimited
Performance:	Surfaced: 24 knots Submerged: 28 knots

Orzel

Orzel was ordered in January 1935 and was funded by public subsciption. She was a large, ocean-going boat with excellent all-round qualities and was Dutch-built, together with her sister ship *Wilk* (Wolf). Diving depth was 80m (200ft) and submerged range was 190km (102nm) at five knots. *Orzel* was commissioned in February 1939. On 14 September 1939 the Polish submarines were ordered to break out from the Baltic and make for British ports; *Wilk* arrived on 20 September and *Orzel* (Lt-Cdr Grudzinski) on 14 October via Reval, after an adventurous voyage without charts. On 8 April 1940 *Orzel* sank two large troop transports at the start of the German invasion of Norway, but was lost in a mine barrage off the Norwegian coast on 8 June. Her sister, *Wilk*, attacked and sank a Dutch submarine in error on 20 June 1940.

Country:	Poland
Launch date:	1938
Crew:	56
Displacement:	Surfaced: 1117 tonnes (1100 tons)
	Submerged: 1496 tonnes (1473 tons)
Dimensions:	84m x 6.7m x 4m (275ft 7in x 22ft 13ft 1in)
Armament:	Twelve 550mm (21.7in) torpedo tubes, One 105mm (4in) gun
Powerplant:	Twin screws, diesel/electric motors
Surface range:	13,300km (7169nm) at 10 knots
Performance:	Surfaced: 15 knots
	Submerged: 8 knots

Oscar class

The underwater equivalent of a Kirov-class battlecruiser, the first Oscar I-class cruise-missile submarine (SSGN) was laid down at Severodvinsk in 1978 and launched in the spring of 1980, starting sea trials later that year. The second was completed in 1982, and a third of the class – which became the first Oscar II – completed in 1985, followed by another ten. The primary task of the Oscar class was to attack NATO carrier battle groups with a variety of submarine-launched cruise missiles, including the SS-N-19 Shipwreck. The SSM tubes are in banks of 12 either side and external to the pressure hull and are inclined at 40°, with one hatch covering each pair. *Kursk*, the 13th boat of the class, was lost in August 2000 after a torpedo being loaded into a tube exploded, starting a fire that caused several more torpedoes to explode. All 118 crew died.

Country:	Russia
Launch date:	April 1980
Crew:	130
Displacement:	Surfaced: 11,685 tonnes (11,500 tons) Submerged: 13,615 tonnes (13,400 tons)
Dimensions:	143m x 18.2m x 9m (469ft 2in x 59ft 8in x 29ft 6in)
Armament:	SS-N-15, SS-N-16 and SS-N-19 SSMs; four 533mm (21in) and four 650mm (25.6in) torpedo tubes
Powerplant:	Two shafts; two nuclear PWR; two turbines
Range:	Unlimited
Performance:	Surfaced: 22 knots Submerged: 30 knots

Oyashio

The first boat in a new class of Japanese SSK, *Oyashio,* was laid down in January 1994. Eleven boats were built, the construction work shared between Mitsubishi and Kawasaki at the Kobe shipyards. Although some damage was caused to the latter by the Kobe earthquake in Kanuary 1995, production was not disrupted, with the last commissioned in 2008. The boats are fitted with large flank sonar arrays; there are anechoic tiles on the fin and double hull sections fore and aft. The class was developed from the earlier Harushio-class boats and the two classes strongly resemble one another, although the Oyashios have a greater displacement. The Oyashio class are being refitted to a similar level to the latest Soryu class submarines. The first two boats, *Oyashio* and *Michishio*, were converted to training vessels between 2015 and 2017.

Country:	Japan
Launch date:	15 October 1996
Crew:	69
Displacement:	Surfaced: 2743 tonnes (2700 tons) Submerged: 3048 tonnes (3000 tons)
Dimensions:	81.7m x 8.9m x 7.9m (268ft x 29ft 2in x 25ft 11in)
Armament:	Six 533mm (21in) torpedo tubes; Sub Harpoon SSM
Powerplant:	Single shaft, diesel/electric motors
Surface range:	Classified
Performance:	Surfaced: 12 knots Submerged: 20 knots

Papa

In 1970 the Soviet shipyard at Severodvinsk launched a single unit of what came to be known in NATO circles as the Papa class. The boat was considerably larger and had two more missile tubes than the contemporary Charlie-class SSGNs, and was for long a puzzle to Western intelligence services. The answer appeared in 1980 at the same shipyard with the even larger Oscar-class SSGN; the Papa-class unit had been the prototype for advanced SSGN concepts with a considerably changed powerplant and a revised screw arangement incorporating five or seven blades. The missile system's function had been to test the underwater-launched version of the SS-N-9 Siren for the subsequent Charlie II series of SSGN. The Oscar design produced yet further improvements, with two 12-round banks of submerged-launch long-range SS-N19 anti-ship missile tubes.

Country:	Russia
Launch date:	1970
Crew:	110
Displacement:	Surfaced: 6198 tonnes (6100 tons) Submerged: 7112 tonnes (7000 tons)
Dimensions:	109m x 11.5m x 7.6m (357ft 7in x 37ft 9in x 24ft 11in)
Armament:	Six 533mm (21in) and two 406mm (16in) torpedo tubes
Powerplant:	Two shafts, one nuclear PWR, two turbines
Range:	Unlimited
Performance:	Surfaced: 20 knots Submerged: 39 knots

Parthian

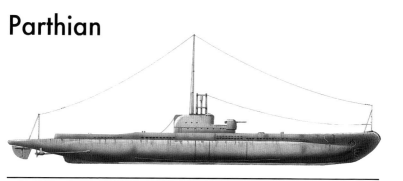

The six vessels in the Parthian class were laid down in 1928 and completed in 1930-31. All were fitted with Vulcan clutches and high capacity batteries. The 14 torpedoes carried were Mk VIIIs, standard armament on all subsequent British submarines of that period. During World War II, the surviving boats of the Parthian class had a 20mm Oerlikon added and could take 18 M2 mines, laid from the torpedo tubes, in place of torpedoes. All were originally deployed in Chinese waters, but transferred to the Mediterranean in 1940. *Parthian* went missing in the Adriatic on 11 August 1943, presumed mined; *Perseus* was torpedoed by the Italian submarine *Enrico Toti* off Zante; *Phoenix* was sunk by the Italian torpedo-boat *Albatros* off Sicily; and *Pandora* was bombed by Italian aircraft at Malta. Only *Proteus* survived, to become a training vessel.

Country:	Britain
Launch date:	22 June 1929
Crew:	53
Displacement:	Surfaced: 1788 tonnes (1760 tons) Submerged: 2072 tonnes (2040 tons)
Dimensions:	88.14m x 9.12m x 4.85m (289ft 2in x 29ft 11in x 15ft 11in)
Armament:	Eight 533mm (21in) torpedo tubes; one 102mm (4in) gun
Powerplant:	Two shafts, diesel/electric motors
Surface range:	9500km (5633nm) at 10 knots
Performance:	Surfaced: 17.5 knots Submerged: 8.6 knots

Pickerel

The Tench class marked the ultimate refinement in the basic design whose ancestry could be traced back to the P class. Externally they were virtually identical to the Balaos. Only a dozen managed to see operational duty in World War II and none of these was lost. Total production was 33 boats between 1944 and 1946, with another 101 cancelled or scrapped incomplete. Differences from the earlier Balaos, though not obvious, were significant. Engine noise was reduced, and the fuel and ballast tanks were better organized. Even a further four torpedo reloads were squeezed in, and this, combined with radar and efficient mechanical fire-control computers put the Tenches way ahead of the opposition. *Pickerel* was transferred to Italy in 1972 after extensive updating, where she served under the name *Gianfranco Gazzana Priaroggia* until 1981.

Country:	USA
Launch date:	15 December 1944
Crew:	22
Displacement:	surfaced: 1595 tonnes (1570 tons) submerged: 2453 tonnes (2415 tons)
Dimensions:	95.2m x 8.31m x 4.65m (311ft 8in x 27ft 3in x 15ft 3in)
Armament:	One or two 127mm (5in) guns;10 533mmTT for 28 topredoes
Powerplant:	Four diesels and two electric motors
Surface range:	20,372km (11,000nm) at 10 knots
Performance:	surfaced: 20.2 knots submerged: 8.7 knots

Pietro Micca

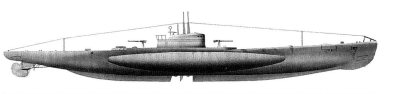

This was a long-range, torpedo and minelaying boat with a partial double hull and an operation depth of 90m (295ft). Although an experimental vessel, and not repeated, she was a good, manoeuvrable seaboat. Surfaced range at full speed was 4185km (2256nm). Her first war operation, on 12 June 1940, under Cdr Meneghini, was to lay a barrage of 40 mines off Alexandria; on 12 August she laid more mines in the same area, and unsuccessfully attacked a destroyer two days later. In February and March 1941 she ran supplies from Tobruk to the Italian garrison on the island of Leros, and in summer 1942 she joined other submarines in transporting fuel and supplies to ports in Cyrenaica. She continued to act as a transport submarine until her career came to an abrupt end on 29 July 1943 when she was sunk in the Straits of Otranto by HM submarine *Trooper*.

Country:	Italy
Launch date:	31 March 1935
Crew:	72
Displacement:	Surfaced: 1595 tonnes (1570 tons) Submerged: 2000 tonnes (1970 tons)
Dimensions:	90.3m x 7.7m x 5.3m (296ft 3in x 25ft 3in x 17ft 4in)
Armament:	Six 533mm (21in) torpedo tubes; two 120mm (4.7in) guns
Powerplant:	Twin screws, diesel/electric motors
Surface range:	10,300km (5552nm) at 9 knots
Performance:	Surfaced: 14.2 knots Submerged: 7.3 knots

Pioneer

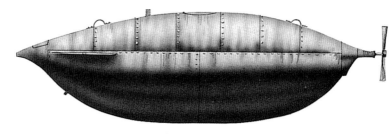

In the American Civil War, the Confederacy was especially interested in any development which promised to break the Union maritime blockade. This early submersible, the only privateer submarine ever built, was laid down in late 1861 at the government yard at New Basin in New Orleans. She had an oval-shaped hull, as did others that succeeded her. Operated by a three–man crew, two of whom worked the hand cranking system which drove the external single propeller, her only armament was a spar torpedo, a long shaft with explosive at the end which had to be rammed into the side of the target vessel. In March 1862 *Pioneer* was issued with a Letter of Marque, which licensed her to sink Union warships. Her crew would collect a bounty of 20 per cent of the estimated value of their victim. In 1952 she was moved to the Louisiana State Museum.

Country:	Confederate States of America
Launch date:	February 1862
Crew:	3
Displacement:	Surfaced: 4 tonnes /tons Submerged: Not known
Dimensions:	10.3m x 1.2m x 1.2m (34ft x 4ft x 4ft)
Armament:	One spar torpedo
Powerplant:	Single screw, hand-cranked
Surface range:	Not known
Performance:	Surfaced: Not known Submerged: Not known

Piper

Piper (SS409) was a double-hull, ocean-going submarine with good seakeeping qualities and range. She was one of the Gato class of over 300 boats, and as such was part of the largest warship project undertaken by the US Navy. These boats were to wreak havoc on Japan's mercantile shipping in the Pacific war. *Piper* was originally named *Awa* and, like many other Gato-class boats, was built at the Portsmouth Naval Dockyard. She deployed to the Pacific late in 1944, and became combat-ready early in the New Year. On 10 February 1945, together with the submarines *Sterlet, Pomfret, Trepang, Bowfin, Sennet, Lagarto* and *Haddock,* she hunted enemy patrol boats that might have detected the presence of Vice-Admiral Mitscher's Task Force 58, which was heading for Iwo Jima. *Piper* was placed in reserve some years after the war and was stricken in 1970.

Country:	USA
Launch date:	26 June 1944
Crew:	80
Displacement:	Surfaced: 1854 tonnes (1825 tons) Submerged: 2448 tonnes (2410 tons)
Dimensions:	95m x 8.3m x 4.6m (311ft 9in x 27ft 3in x 15ft 3in)
Armament:	Ten 533mm (21in) torpedo tubes; one 76mm (3in) gun
Powerplant:	Twin screw diesels, electric motors
Surface range:	22,236km (12,000nm) at 10 knots
Performance:	Surfaced: 20 knots Submerged: 10 knots

Porpoise

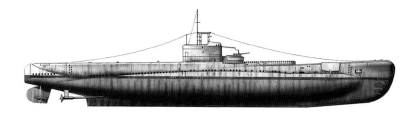

HMS *Porpoise* was leader of a class of six boats launched between 1932 and 1938. Three further boats were cancelled in 1941. The vessels of the class served in many different theatres from home waters to the West Indies, the Mediterranean, and the China Station, and five were lost in various ways, with *Rorqual,* deployed to the Eastern Fleet in 1944, the only one to survive the Second World War. *Grampus* was sunk by the Italian torpedo-boats *Cl10* and *Circe* off Augusta on 24 June 1940; *Narwhal* went missing off Norway in July 1940; *Cachalot* was rammed by the Italian torpedo-boat *Papa* off Cyrenaica on 4 August 1941; *Seal* was damaged by a mine before surrendering to the Germans on 5 May 1940; and *Porpoise* herself was bombed and sunk by Japanese aircraft in the Malacca Strait on 19 January 1945.

Country:	Britain
Launch date:	30 August 1932
Crew:	61
Displacement:	Surfaced: 1524 tonnes (1500 tons) Submerged: 2086 tonnes (2053 tons)
Dimensions:	81.5m x 9m x 13.75m (267ft 9in x 29ft 9in x 13ft 9in)
Armament:	Six 533mm (21in) torpedo tubes; one 102mm (4in) gun
Powerplant:	Twin screws, diesel/electric motors
Surface range:	10,191km (5500nm) at 10 knots
Performance:	Surfaced: 15 knots Submerged: 8.75 knots

R1

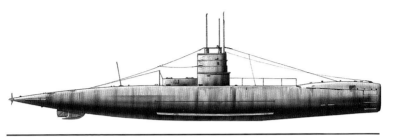

In 1917, the Royal Navy was struggling to counter the German U-boats' toll of British merchant shipping. The most significant development was the convoy system, but proposals for submarines able to hunt other submarines were also considered. *R1* was the forerunner of the modern ASW hunter-killer fleet submarine. Ten vessels were completed. They were intended to chase U-boats on the surface, and sink them with torpedoes. *R1* had a hull form similar to that of the earlier H class, and a pronounced bulge above the bow. She was highly streamlined, with internal ballast tanks and no gun. Five powerful hydrophones with bearing instruments were carried in the bow compartment. These vessels were a daring solution to a problem which nearly brought Britain to disaster, but they arrived too late to have any effect.

Country:	Britain
Launch date:	April 1918
Crew:	36
Displacement:	Surfaced: 416 tonnes (410 tons) Submerged: 511 tonnes (503 tons)
Dimensions:	49.9m x 4.6m x 3.5m (163ft 9in x 15ft x 11ft 6in)
Armament:	Six 457mm (18in) torpedo tubes
Powerplant:	Single screw, diesel/electric 1200 hp motors
Surface range:	3800km (2048nm) at 8 knots
Performance:	Surfaced: 15 knots Submerged: 9.5 knots

Redoutable

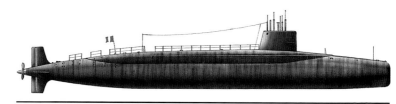

Laid down at Cherbourg Naval Dockyard on 30 March 1964, *Le Redoutable* was France's first ballistic-missile submarine, and the prototype of the seaborne element of the French Force de Dissuasion (Deterrent Force), which ultimately consisted of boats armed with the MSBS (Mer-Sol Ballistique Strategique) IRBM. The French term for SSBN is SNLE (Sous-marin Nucleaire Lance-Engins). *Le Redoutable* reached IOC in December 1971; *Le Terrible* followed in 1973, *Le Foudroyant* in 1974, *L'Indomptable* in 1977 and *Le Tonnant* in 1979. Later, all units except *Le Redoutable* were armed with the Aerospatiale M4 three-stage solid fuel missile, which has a range of 5300km (2860nm) and carries six MIRV, each of 150kT. The M4 missiles could be discharged at twice the rate of the M20. *Le Redoutable* was withdrawn in 1991.

Country:	France
Launch date:	29 March 1967
Crew:	142
Displacement:	Surfaced: 7620 tonnes (7500 tons)
	Submerged: 9144 tons (9000 tons)
Dimensions:	128m x 10.6m x 10m (420ft x 34ft 10in x 32ft 10in)
Armament:	16 submarine-launched MRBMs
Powerplant:	One nuclear PWR, turbines
Range:	Unlimited
Performance:	Surfaced: 20 knots
	Submerged: 28 knots

Reginaldo Giuliani

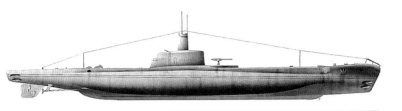

Reginaldo Giuliani and her three sisters of the Liuzzi class were developments of the Brin class. Maximum diving depth was 90m (290ft). In 1940 *Giuliani* was deployed to the French Atlantic ports for operations against British convoys; she was later converted to transport cargo to the Far East. Seized by the Japanese at Singapore after Italy's surrender, she was handed over to Germany as *UIT23*. On 14 February 1944 she was intercepted and sunk in the Malacca Straits by HMS *Tally Ho* (Cdr Bennington). A sister ship, *Alpino Bagnolini*, also captured, and renumbered *UIT 22*, was sunk by South African aircraft off the Cape of Good Hope on 11 March 1944; another, *Capitano Tarantini,* was sunk on 15 December 1940 by HM submarine *Thunderbolt. Console Generale Liuzzo* was scuttled off Crete after a battle with the destroyers HMS *Defender, Dainty* and *Ilex.*

Country:	Italy
Launch date:	30 December 1939
Crew:	58
Displacement:	Surfaced: 1184 tonnes (1166 tons) Submerged: 1507 tonnes (1484 tons)
Dimensions:	76.5m x 6.8m x 4.7m (251ft x 22ft 4in x 15ft 5in)
Armament:	Eight 533mm (21in) torpedo tubes; one 100mm (3.9in) gun
Powerplant:	Twin screw, diesel/electric motors
Surface range:	19,950km (10,750nm) at 8 knots
Performance:	Surfaced: 17.5 knots Submerged: 8.4 knots

Remo

Remo was leader of the R class of 12 transport submarines, the largest Italian boats built up to that time. Laid down in September 1942, she was completed in June 1943. She had four watertight holds with a total capacity of 600 cubic metres (21,190 cubic feet). Maximum diving depth was 100m (328ft). She and her one operational sister, *Romolo*, were developed to transport cargo to and from the Far East. Ten more vessels in the class were laid down; two (*R3* and *R4*) were launched in 1946 and then scrapped, two (*R5* and *R6*) were scrapped on the slip, and the rest were either scuttled or sunk in air attacks after being seized by German forces. *R11* and *R12* were subsequently refloated and used as oil storage vessels for some years after the war. *Remo* was sunk by HM submarine *United* on her maiden voyage in the Gulf of Taranto on 15 July 1943..

Country:	Italy
Launch date:	28 March 1943
Crew:	63
Displacement:	Surfaced: 2245 tonnes (2210 tons)
	Submerged: 2648 tonnes (2606 tons)
Dimensions:	70.7m x 7.8m x 5.3m (232ft x 25ft 9in x 17ft 6in)
Armament:	Three 20mm (0.8in) guns
Powerplant:	Twin screws, diesel/electric motors
Surface range:	22,236km (12,000nm) at 9 knots
Performance:	Surfaced: 13 knots
	Submerged: 6 knots

Requin

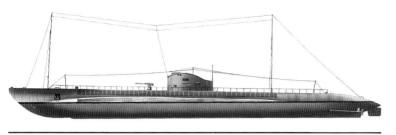

L aid down in 1923 and completed three years later, *Requin* (Shark) was leader of a class of nine submarines that were completely modernized in 1935-37. They had chequered fortunes during World War II. *Caiman* and *Marsouin* escaped from Algiers in November 1942, at the time of the Allied landings in North Africa and sailed to Toulon, where *Caiman* was scuttled; *Requin, Dauphin, Phoque* and *Espadon* were seized by the Germans at Bizerta in December 1942 but were never commissioned, being broken up later. *Morse* struck a mine and sank off Sfax on 10 June 1940; *Souffleur* was sunk by the British submarine *Parthian* (Cdr Rimington) during the Syrian campaign on 29 June 1941; and *Narval* was sunk by a mine on 15 December 1940, en route for Malta, her captain having refused to return to Toulon after France's surrender.

Country:	France
Launch date:	19 July 1924
Crew:	54
Displacement:	Surfaced: 974 tonnes (990 tons) Submerged: 1464 tonnes (1441 tons)
Dimensions:	78.25m x 6.84m x 5.10m (256ft 7in x 22ft 6in x 16ft 9in)
Armament:	Ten 550mm (21.7in) torpedo tubes
Powerplant:	Twin screws, diesel/electric motors
Surface range:	10,469km (5650nm) at 10 knots
Performance:	Surfaced: 15 knots Submerged: 9 knots

Requin

Requin was one of six Narval-class diesel-electric boats laid down for the French Navy between 1951 and 1956. The requirement was for a submarine of 1219 tonnes (1200 tons) standard displacement with a surfaced speed of 16 knots and a range of 27,795km (15,000nm) with snorkel. France still had substantial colonial interests in the Pacific and Indo-China, making it imperative for the new class of submarine to be capable of a fast transit over long distances followed by a patrol of seven to fourteen days. The Narvals were in fact improved versions of the German Type XXI, one of which had been extensively tested after the war under the name *Roland Morillot*. The other boats in the class were *Dauphin* (Dolphin), *Espadon* (Swordfish), *Marsouin* (Porpoise), *Morse* (Walrus) and *Narval* (Narwhal) – names taken from a pre-war class of French submarine.

Country:	France
Launch date:	3 December 1955
Crew:	63
Displacement:	Surfaced: 1661 tonnes (1635 tons)
	Submerged: 1941 tonnes (1910 tons)
Dimensions:	78.4m x 7.8m x 5.2m (257ft x 26ft x 17ft)
Armament:	Eight 550mm (21.7in) torpedo tubes
Powerplant:	Two shafts, diesel/electric motors
Surface range:	27,795km (15,000nm) at 8 knots
Performance:	Surfaced: 16 knots
	Submerged: 18 knots

Resolution

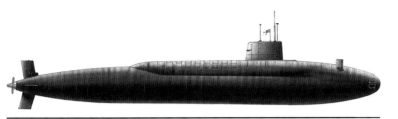

In February 1963 the British Government stated its intention to order four or five Resolution-class nuclear-powered ballistic missile submarines, armed with the American Polaris SLBM, to take over the British nuclear deterrent role from the RAF from 1968. With characteristics very similar to the American Lafayettes, the lead ship HMS *Resolution* was commissioned in October 1967. HMS *Repulse* followed in September 1968, with the HMS *Renown* and HMS *Revenge* commissioning in November 1968 and December 1969 respectively. Early in 1968 *Resolution* undertook missile trials with Polaris off Florida, and four months later she made her first operational patrol. In the 1990s the Resolution-class boats were progressively replaced by the new Vanguard class SSBNs, armed with the Trident II missile. The first of these was commissioned in August 1993.

Country:	Britain
Launch date:	September 1966
Crew:	154
Displacement:	Surfaced: 7620 tonnes (7500 tons) Submerged: 8535 tonnes (8400 tons)
Dimensions:	129.5m x 10.1m x 9.1m (425ft x 33ft x 30ft)
Armament:	Sixteen Polaris A3TK IRBMs; six 533mm (21in) torpedo tubes
Powerplant:	Single shaft, one nuclear PWR, two steam turbines
Range:	Unlimited
Performance:	Surfaced: 20 knots Submerged: 25 knots

Resurgam II

Resurgam II was designed by George Garrett. She had a cigar-shaped hull with spindle ends, and was able to withstand a pressure of 71 pounds per square inch (32kg per square mm), enabling her to dive to a depth of about 45.5m (150ft). Propulsion was by steam on the Lamm fireless principle. A coal furnace heated water in a large steam boiler; the fire door and smoke-escape valve leading to a short funnel inside the superstructure were then closed. Latent heat turned the water into steam when the throttle valve was opened, thereby supplying the engine. Initial trials were promising, and Garrett decided to set up base on the Welsh coast. However, as Resurgam II was being towed to her new berth in February 1880, she sank during a storm. She still lies where she sank off the British coast, although there are plans to raise her.

Country:	Britain
Launch date:	December 1879
Crew:	2
Displacement:	Surfaced: 30 tonnes/tons Submerged: Not known
Dimensions:	13.7m x 2.1m (45ft x 7ft)
Armament:	None
Powerplant:	Single screw, Lamm steam locomotive
Surface range:	Not known
Performance:	Surfaced: 3 knots Submerged: Not known

RO100

Ordered under the Japanese Navy's 1940 and 1941 programmes, the RO100 (Type KS) class were designed as coastal submarines to be used around the Japanese coastline and in the waters of the outposts of the Japanese Empire. Designed to operate near their bases, their operational endurance was only 21 days. Submerged range was 60 nautical miles at 3 knots, and they had a diving depth of 75m (245ft). Production was shared between the Kure Navy Yard and the Kawasaki Yard at Kobe. None of the class survived the war, during which they sank six merchant ships totalling 35,247 tonnes (34,690 tons) and damaged three more totalling 14,300 tonnes (14,074 tons). In addition, *RO106* (Lt Nakamura) sank the tank landing ship *LST 342* off New Georgia on 18 July 1943, while *RO108* sank the US destroyer *Henley* off Finschhafen (New Guinea) on 3 October 1943.

Country:	Japan
Launch date:	1942
Crew:	75
Displacement:	Surfaced: 611 tonnes (601 tons) Submerged: 795 tonnes (782 tons)
Dimensions:	57.4m x 6.1m x 3.5m (188ft 3in x 20ft x 11ft 6in)
Armament:	Four 533mm (21in) torpedo tubes; one 76mm (3in) gun
Powerplant:	Two shaft diesels plus electric motors
Surface range:	6485km (3500nm) at 12 knots
Performance:	Surfaced: 14 knots Submerged: 8 knots

Roland Morillot

This was the former German Type XXI U-boat *U2518*, one of a group of Horten-based submarines that surrendered to the British in May 1945. She was transferred to France in 1946, and renamed in the following year. The Type XXI was an ocean-going submarine capable of fully-submerged operations with the use of Snorkel apparatus. A conventional diesel-electric boat, it had a streamlined hull of all-welded construction which, in order to speed production, was prefabricated in eight sections. The Type XXI was equipped with chin sonar; it could carry a very useful load of 23 torpedoes, or 12 torpedoes plus 12 mines. France tested the vessel exhaustively, and the lessons learned were incorporated in the *Requin*-class submarines of the 1950s. *Roland Morillot* was stricken from the French navy list in 1968.

Country:	France
Launch date:	1944
Crew:	57
Displacement:	Surfaced: 1638 tonnes (1612 tons) Submerged: 1848 tonnes (1819 tons)
Dimensions:	76.5m x 7m x 6m (251ft x 23ft x 19ft 8in)
Armament:	Six 533mm (21in) torpedo tubes
Powerplant:	Single shaft, diesel/electric motors
Surface range:	17,933km (9678nm) at 12 knots
Performance:	Surfaced: 15.5 knots Submerged: 16 knots

Romeo

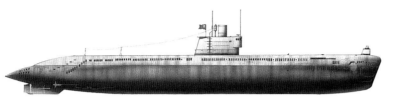

Although it was the Russians who built the first Romeo-class submarines in 1958 at Gorky, as an improvement on the Whiskey design, their construction coincided with the successful introduction of nuclear propulsion into Soviet submarines, so only 20 were completed out of the 560 boats originally planned. The design was passed to the Chinese and production began in China in 1962, at the Jiangnan (Shanghai) shipyard under the local designation Type 033. Three further shipyards then joined to give a maximum yearly production rate of nine units during the early 1970s. Production was completed in 1984 with a total of 98 built for the Chinese Navy; four more were exported to Egypt and seven to North Korea, with a further thirteen built locally with Chinese assistance. In February 1985 the North Koreans lost one of their Romeos with all hands in the Yellow Sea.

Country:	China
Launch date:	1962
Crew:	60
Displacement:	Surfaced: 1351 tonnes (1330 tons) Submerged: 1727 tonnes (1700 tons)
Dimensions:	77m x 6.7m x 4.9m (252ft 7in x 22ft x 16ft 1in)
Armament:	Eight 533mm (21in) torpedo tubes
Powerplant:	Twin screws, diesel/electric motors
Surface range:	29,632km (16,000nm) at 10 knots
Performance:	Surfaced: 16 knots Submerged: 13 knots

Rubis

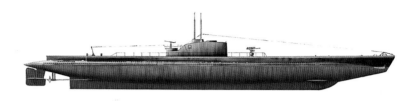

Rubis was one of six vessels of the Saphir class of minelaying submarine. Of all French submarines, these were arguably the best suited to their task. Their mines were stowed in wells in the outer ballast tanks and had direct release mechanism. In March 1940 *Rubis* joined the 10th Flotilla and the 2nd Submarine Division at Harwich, and began minelaying operations off Norway on 10 May, claiming eleven enemy freighters and the submarine-chaser *UJD*. The collapse of France in June 1940 found her in Dundee, and her crew elected to join the Free French forces, continuing to operate in the North Sea. In August 1941 she torpedoed a freighter off Norway. Minelaying later in the war claimed at least four more feighters, as well as the submarine chasers *UK1113*, *UJ1116* and *R402* on 21 December 1944, making her the most successful minelayer of the war.

Country:	France
Launch date:	30 September 1931
Crew:	42
Displacement:	Surfaced: 773 tonnes (761 tons) Submerged: 940 tonnes (925 tons)
Dimensions:	65.9m x 7.12m x 4.3m (216ft 1in x 23ft 3in x 14ft 6in)
Armament:	Three 550mm (21.7in) and two 400mm (15.7in) torpedo tubes; one 75mm (3in) gun; 32 mines
Powerplant:	Twin screws, diesel/electric motors
Surface range:	12,971km (7000nm) at 7.5 knots
Performance:	Surfaced: 12 knots Submerged: 9 knots

Rubis

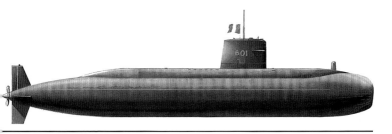

The Rubis class of Fleet Nuclear-Attack Submarines (Sous-Marins Nucléaires d'Attaque, or SNA) comprises eight vessels, in two squadrons, one based at Lorient to cover the SSBN base and the other at Toulon. Eight boats were built: *Rubis* (S601), *Saphir* (S602), *Casabianca* (S603), *Emeraude* (S604), *Améthyste* (S605), *Perle* (S606), *Turquoise* (S607) and *Diamant* (S608). The last five were built to a modified design, including a new bow form and silencing system, as well as new tactical and attack systems and improved electronics. Apart from the experimental 406-tonne (400-ton) *NR-1* of the US Navy, *Saphir* is the smallest nuclear-attack submarine ever built, reflecting France's advanced nuclear-reactor technology. *Rubis* became operational in February 1983, the last, *Diamant,* in 1999. The Suffren class submarines begin to replace them in service in 2020.

Country:	France
Launch date:	7 July 1979
Crew:	67
Displacement:	Surfaced: 2423 tonnes (2385 tons) Submerged: 2713 tonnes (2670 tons)
Dimensions:	72.1m x 7.6m x 6.4m (236ft 6in x 24ft 11in x 21ft)
Armament:	Four 533mm (21in) torpedo tubes; Exocet SSMs
Powerplant:	Single shaft, one nuclear PWR, auxiliary diesel/electric
Range:	Unlimited
Performance:	Surfaced: 25 knots Submerged: Classified

S1

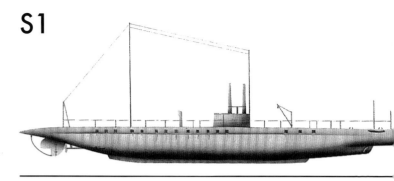

$S1$ was based on the Italian Laurenti type, having a partial double hull with ten watertight compartments. Her diesel engines developed 650hp, the electric motors 400hp. Only three boats of the S class were completed. There were no fewer than 23 classes of submarine in service with the Royal Navy during World War I, of which by far the most numerous was the E class, with 58 boats. The C class came next, with 38, followed by the H1 class with 20, the K class with 17, the G class with 14, the A class with 13, the B class with 11 and the H21 and K classes with 10 each. The building of some classes continued after the war; 14 more H21s were built after 1919, as were 11 of the L9 class, which had eight boats in service during World War I. All three S-class boats, $S1$, $S2$ and $S3$, were transferred to the Italian Navy in 1915, and were discarded in 1919.

Country:	Britain
Launch date:	28 February 1914
Crew:	31
Displacement:	Surfaced: 270 tonnes (265 tons) Submerged: 330 tonnes (324 tons)
Dimensions:	45m x 4.4m x 3.2m (148ft x 14ft 5in x 10ft 6in)
Armament:	Two 457mm (18in) torpedo tubes, one 12-pounder gun
Powerplant:	Twin screws, diesel/electric motors
Surface range:	2963km (1600nm) at 8.5 knots
Performance:	Surfaced: 13 knots Submerged: 8.5 knots

S28

In December 1941, more than half the USA's submarines belonged to the old O, R and S classes of World War I vintage. The so-called Old S class comprised four separate groups of 38 boats. In October 1943 *S28*, under Lt-Cdr Sislet, sank a Japanese freighter in the Pacific, but no further successes were recorded before she herself was lost in July 1944, having failed to surface during a training exercise at Pearl Harbor. Some of the S class were transferred to the Royal Navy in 1942; *S1* became *P552, S21 P553, S22 P554, S24 P555, S25 P551* and then the Polish *Jastrzab*, and *S29 P556*. War losses among the remainder included *S26* (sunk in a collision on 24 January 1942), *S27* (ran aground on 19 June 1942), *S36* (ran aground on 20 January 1942), *S39* (ran aground on 14 August 1942), and *S44* (sunk by Japanese destroyer, 7 October 1943).

Country:	USA
Launch date:	20 September 1922
Crew:	42
Displacement:	Surfaced: 864 tonnes (850 tons)
	Submerged: 1107 tonnes (1090 tons)
Dimensions:	64.3m x 6.25m x 4.6m (211ft x 20ft 6in x 15ft 3in)
Armament:	Four 533mm (21in) torpedo tubes; one 102mm (4in) gun
Powerplant:	Two-shaft diesels
Surface range:	6333km (3420nm) at 6.5 knots
Performance:	Surfaced: 14.5 knots
	Submerged: 11 knots

San Francisco

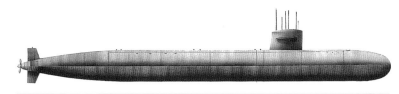

One of the Los Angeles class originally designed as a counter to the Russian Victor-class SSN, *San Francisco* was the third vessel built and launched in 1979. On 8 January 2005 she collided with a seamount in the Pacific Ocean while travelling at flank speed at a depth of 160m (525ft). Her bow was crushed, rupturing the forward ballast tanks, but her pressure hull and reactor were not damaged. Ninety-eight crew were injured, one later dying of his wounds. The crew managed to surface the submarine, and she limped to back to Guam. The captain was relieved of his command, and six other crew were reduced in rank after the incident. As she had been recently refuelled, it was decided to replace her bow with that of her sister, *Honolulu*, as it was cheaper to repair *San Francisco* than refuel *Honolulu*. She was finally decommissioned in 2022.

Country:	USA
Launch date:	27 October 1979
Crew:	133
Displacement:	Surfaced: 6180 tonnes (6082 tons) Submerged: 7038 tonnes (6927 tons)
Dimensions:	110.3m x 10.1m x 9.9m (362ft x 33ft x 32ft 3in)
Armament:	Four 533mm (21in) torpedo tubes; Tomahawk Land Attack Cruise Missiles, Harpoon SSM
Powerplant:	Single shaft, nuclear PWR, turbines
Range:	Unlimited
Performance:	Surfaced: 20 knots Submerged: 32 knots

Sanguine

This was one of the last batch of S-class boats built by Cammell Laird and laid down in 1944–45. The class was not scheduled for any major modification post-war, but several vessels were modified for trials work. Israel bought two, *Sanguine* and *Springer*, in 1958, and named them *Rahav* and *Tanin* respectively. The latter was discarded as worn out in 1968 and cannibalised to keep *Rahav* in commission for some time longer in the training role. They were replaced in Israeli service by HM submarines *Turpin* and *Truncheon*, which were commissioned into the Israeli Navy as *Leviathan* and *Dolphin*. Three S-class boats, *Spur*, *Saga* and S*pearhead* were transferred to Portugal in 1948–49, while the quartet of *Styr*, *Spiteful*, *Sportsman* and *Statesman* were transferred to France in 1951–52.

Country:	Britain
Launch date:	15 February 1945
Crew:	44
Displacement:	Surfaced: 726 tonnes (715 tons)
	Submerged: 1006 tonnes (990 tons)
Dimensions:	61.8m x 7.25m x 3.2m (202ft 6in x 23ft 9in x 10ft 6in)
Armament:	Six 533mm (21in) torpedo tubes; one 76mm (3in) gun
Powerplant:	Twin screws, diesel/electric motors
Surface range:	15,750km (8500nm) at 10 knots
Performance:	Surfaced: 14.7 knots
	Submerged: 9 knots

Santa Cruz

The long-range HY80 steel-hulled *Santa Cruz* or TR1700 class of SSK was designed by the German firm of Thyssen Nordseewerk to meet Argentine Navy requirements. Ordered in 1977, the first two units, *Santa Cruz* and *San Juan*, were built at Emden and commissioned into the Argentine Navy in 1984 and 1985 respectively. The contract specified that four more boats were to be built in Argentina with German assistance; two, *S43* (*Santa Fe*) and *S44* (*Santiago del Esturo*), were laid down, but construction was suspended due to the Argentine economic crisis. Parts from the new boats were used as spares. *San Juan* was lost with all 44 hands in November 2017. A major search effort failed to locate the wreck; it was only found a year later. *Santa Cruz* was undergoing a refit, which was paused then cancelled. She remains inoperative.

atCountry:	Argentina
Launch date:	28 September 1982
Crew:	29
Displacement:	Surfaced: 2150 tonnes (2116 tons)
	Submerged: 2300 tonnes (2264 tons)
Dimensions:	66m x 7.3m x 6.5m (216ft 6in x 23ft 11in x 21ft 3in)
Armament:	Six 533mm (21in) torpedo tubes
Powerplant:	Single shaft, diesel/electric motors
Surface range:	22,224km (12,000nm) at 8 knots
Performance:	Surfaced: 15 knots
	Submerged: 25 knots

Scorpène

Originally developed by the French DCNS and Spain's Navantia (before Spain cancelled their order in favour of the S-80), the Scorpène class is available in different configurations and currently operated by several countries, including Chile, Malaysia, India, and Brazil. Chile's navy operates two boats, the *O'Higgins* and *Carrera*, which replaced two Oberon-class submarines. Malaysia also has two, *Tunku Abdul Rahman* and *Tun Abdul Razak*, both commissioned in 2009. India committed to build six submarines in 2005, with the first, INS *Kalvari*, commissioned in 2017; the sixth, *Vagsheer*, will be commissioned in 2024. Brazil built four enlarged Scorpènes, known as the Riachuelo class, with *Riachuelo* and *Humaitá* commissioned in 2022 and 2024, respectively. Indonesia signed a contract to build two Scorpène Evolved vessels in April 2024.

Country:	France
Launch date:	1 November 2003
Crew:	31
Displacement:	Surfaced: 1870 tonnes (1840 tons)
	Submerged: 2032 tonnes (2000 tons)
Dimensions:	70m x 6.2m x 5.8m (230ft x 20ft 4in x 19ft)
Armament:	Six 533mm (21in) tubes for 18 torpedoes or SM39 Exocet
	anti-ship missiles or 30 mines
Powerplant:	Two diesels, electric motor; AIP propulsion unit option; single screw
Surface Range:	12,038km (6500nm) at 8 knots
Performance:	Surfaced: 12 knots
	Submerged: 20 knots

Seawolf

Originally conceived during the Cold War era as a replacement for the Los Angeles-class submarines, the Seawolf class was designed to be exceptionally quiet, fast, and heavily armed, making them highly capable of conducting covert missions and hunting enemy submarines. However, due to budget constraints (each Seawolf cost around $3bn) and changing strategic priorities following the end of the Cold War, the class was truncated to just three submarines instead of the planned 29, and the smaller Virginia class was built instead. The three built were: USS *Seawolf* (SSN-21), USS *Connecticut* (SSN-22), and USS *Jimmy Carter* (SSN-23). *Jimmy Carter* was lengthened by 30m (100ft) to include a Multi-Mission Platform allowing the use of remotely operated underwater vehicles and Navy SEALS, as well as the ability to tap underwater communications cables.

Country:	USA
Launch date:	24 June 1995
Crew:	140
Displacement:	Surfaced: 8738 tonnes (8600 tons) Submerged: 9285 tonnes (9138 tons)
Dimensions:	107m x 12m x 10.9m (351ft x 40ft x 35ft 9in)
Armament:	Eight 660mm (25.9in) launch tubes for Mk 48 torpedoes, Tomahawk or Sub- Harpoon missiles; 50 torpedoes
Powerplant:	Nuclear PWR, two turbines, propulsor
Surface range:	Unlimited
Performance:	Surfaced: 18 knots Submerged: 35 knots

Severodvinsk

Intended as a replacement for the Soviet-era Akula class, the Yasen class are nuclear-powered SSGNs intended to conduct a wide range of missions. Construction began in the 1990s, but due to budgetary issues and priority being given to the Borei class SSBNs, the first, *Severodvinsk* (K-329), was not commissioned until 2014. In October 2021 she successfully test-launched two Zircon hypersonic missiles in the White Sea. Twelve are planned to be built in the class; the others built to the upgraded Yasen-M standard are *Kazan*, *Novosibirsk*, *Krasnoyarsk* (all currently operational), with *Arkhangelsk*, *Perm*, *Ulyanovsk*, *Voronezh*, *Vladivostok*, *Bratsk* and two yet to be named still to enter service as of 2024. They are equipped with a modern reactor design that has a core life of over 25 years and is significantly quieter than earlier Russian designs.

Country:	Russia
Launch date:	15 June 2010
Crew:	85 (64 Yasen-M)
Displacement:	Surfaced: 8737 tonnes (8600 tons)
	Submerged: 14,021 tonnes (13,800 tons)
Dimensions:	119m x 13.5m x 8.4m (390ft 5in x 44ft 3in x 27ft 6in)
Armament:	Eight silos for VSL weapons including Zircon, Oniks or Kalibr
	SLBMs, and ten 533mm (21in) torpedo tubes
Powerplant:	Nuclear PWR, single shaft
Surface range:	Unlimited
Performance:	Surfaced: 20 knots
	Submerged: 35 knots (28 knots silent running)

Shark

The USS *Shark* (SSN 591) was one of the five Skipjack class SSNs built in the late 1950s. Until the advent of the Los Angeles class, they were the fastest submarines available to the US Navy, and a principal factor in the potentially deadly game of cat and mouse played by NATO and Warsaw Pact submariners for nearly three decades. The development of nuclear-attack submarines in the United States and USSR began at about the same time, but the designs followed different paths, the Americans concentrating on anti-submarine warfare (ASW) and the Russians on a multi-mission role, encompassing both ASW and surface attack with large anti-ship cruise missiles. Later on the Americans also adopted a multi-mission capability with the deployment of submarine-launched weapons like Sub-Harpoon and Tomahawk, designed for anti-ship and land attack.

Country:	USA
Launch date:	16 March 1960
Crew:	106-114
Displacement:	Surfaced: 3124 tonnes (3075 tons) Submerged: 3556 tonnes (3500 tons)
Dimensions:	76.7m x 9.6m x 8.5m (251ft 9in x 31ft 6in x 27ft 10in)
Armament:	Six 533mm (21in) torpedo tubes
Powerplant:	Single shaft, one nuclear PWR, two steam turbines
Range:	Unlimited
Performance:	Surfaced: 18 knots Submerged: 30 knots

Sierra class

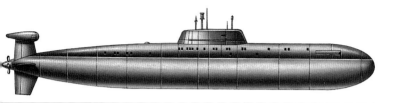

Two SSNs of the Sierra I class, *Carp* and *Kostroma*, were laid down at Gorky and Severodvinsk Shipyards in 1979 and 1984, and commissioned in September 1984 and November 1987. Both were active with Northern Fleet but are now laid up and on the reserve list. *Kostroma* was damaged after colliding with the USS *Baton Rouge* in the Barents Sea in February 1992. Sierra I, known to the Russians as Project 945 *Barrakuda*, was augmented by two vessels of the upgraded Sierra II class, *Pskov* and *Nizhni-Novgorod*. The former was launched in July 1989, the latter in July 1992. The Sierra II (Type 945A *Kondor* class) boats have a diving depth of 750m (2460ft). One notable feature of the Sierras is their titanium hulls allowing them to dive to great depths, titanium being lighter yet stronger than steel, albeit much more expensive and labour-intensive.

Country:	Russia
Launch date:	July 1986 (*Carp*, Sierra I class)
Crew:	61
Displacement:	Surfaced: 7112 tonnes (7000 tons) Submerged: 8230 tonnes (8100 tons)
Dimensions:	107m x 12.5m x 8.8m (351ft x 41ft x 28ft 11in)
Armament:	Four 650mm (25.6in) and four 533mm (21in) torpedo tubeses; SS-N-15 Starfish and SS-N-21 Samson SSMs
Powerplant:	Single shaft, one nuclear PWR, one turbine
Range:	Unlimited
Performance:	Surfaced: 10 knots Submerged: 32 knots

Siroco

The Spanish Navy ordered its first two Agosta-class boats (*Galerna* and *Siroco*) in May 1975, and a second pair *(Mistral* and *Tramontana)* in June 1977. Designed by the French Directorate of Naval Construction as very quiet but high-performance ocean-going diesel-electric boats (SSKs), the Agosta-class boats are each armed with four bow torpedo tubes which are equipped with a rapid-reload pneumatic ramming system that can launch weapons with a minimum of noise signature. The tubes were of a completely new design when the Agostas were authorized in the mid-1970s, allowing a submarine to fire its weapons at all speeds and at any depth down to its maximum operational limit, which in the case of the Agostas is 350m (1148ft). The Spanish Agostas were built with some French assistance, and upgraded in the mid-1990s. *Siroco* was stricken in 2012.

Country:	Spain
Launch date:	13 November 1982
Crew:	54
Displacement:	Surfaced: 1514 tonnes ((1490 tons) Submerged: 1768 tonnes (1740 tons)
Dimensions:	67.6m x 6.8m x 5.4m (221ft 9in x 22ft 4 in x 17ft 9in)
Armament:	Four 550mm (21.7in) torpedo tubes; 40 mines
Powerplant:	Two diesels, one electric motor
Surface range:	Not known
Performance:	Surfaced: 12.5 knots Submerged:17.5 knots

Sjoormen

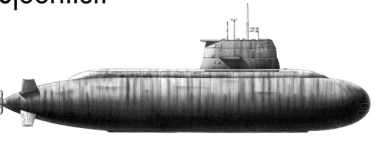

The six boats of the Sjoormen class were the first modern-type submarines to enter service with the Royal Swedish Navy. They were designed in the early 1960s and construction was equally divided between Kockums of Malmø (the designer) and Karlskrona Varvet. With an Albacore-type hull for speed and a twin-deck arrangement, the class was extensively used in the relatively shallow Baltic, where its excellent manoeuvrability and silent-running capabilities greatly enhanced the Swedish Navy's ASW operations. All six boats were upgraded in 1984-85 with new Ericsson IBS-A17 combat data/fire control systems. In the 1990s they were progressively replaced by the new A19 class, which incorporate a fully integrated combat system, more extensive sensors and even quieter machinery to allow their use on offensive (hunter-killer) ASW patrols.

Country:	Sweden
Launch date:	25 January 1967
Crew:	18
Displacement:	Surfaced: 1143 tonnes (1125 tons)
	Submerged: 1422 tonnes (1400 tons)
Dimensions	51m x 6.1m x 5.8m (167ft 3in x 20ft x 19ft)
Armament:	Four 533mm (21in) and two 400mm (15.75in) torpedo tubes
Powerplant:	Single shaft, four diesels, one electric motor
Surface range:	Not known
Performance:	Surfaced: 15 knots
	Submerged: 20 knots

Skate

L aid down in July 1955, the USS *Skate* was the world's first production-model nuclear submarine, followed by three more boats of her class, *Swordfish*, *Sargo* and *Seadragon*. *Skate* made the first completely submerged Atlantic crossing. In 1958 she established a (then) record of 31 days submerged with a sealed atmosphere; on 11 August 1958 she passed under the North Pole during an Arctic cruise; and on 17 March 1959 she became the first submarine to surface at the North Pole. Other boats of the class also achieved notable 'firsts'; in August 1960, *Seadragon* made a transit from the Atlantic to the Pacific via the Northwest Passage (Lancaster Sound, Barrow and McClure Straits). In August 1962 *Skate*, operating from New London, Connecticut, and *Seadragon*, based at Pearl Harbor, made rendezvous under the North Pole.

Country:	USA
Launch date:	16 May 1967
Crew:	95
Displacement:	Surfaced: 2611 tonnes (2570 tons) Submerged: 2907 tonnes (2861 tons)
Dimensions:	81.5m x 7.6m x 6.4m (267 ft 8in x 25ft x 21ft)
Armament:	Six 533mm (21in) torpedo tubes
Powerplant:	Two shafts, one nuclear PWR, turbines
Range:	Unlimited
Performance:	Surfaced: 20 knots Submerged: 25 knots

Skipjack

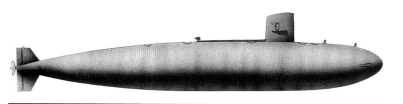

The USS *Skipjack* (SSN 585) was class leader of a group of six nuclear attack submarines built in the late 1950s. The other members of the class were *Scamp* (SSN 588), *Scorpion* (SSN 589), *Sculpin* (SSN 590), *Shark* (SSN 591) and *Snook* (SSN 592). In May 1968, *Scorpion* was lost with all 99 crew members on board some 740km (400nm) southwest of the Azores while en route from the Mediterranean to her base at Norfolk, Virginia. The original *Scorpion* was renumbered SSBN 598 and built as the nuclear ballistic-missile submarine *George Washington*. Unti the advent of the Los Angeles class, the Skipjacks were the fastest submarines available to the US Navy, and had a crucial role to play in the detection, pursuit and destruction of missile submarines from an opposing fleet.

Country:	USA
Launch date:	26 May 1958
Crew:	106-114
Displacement:	Surfaced: 3124 tonnes (3075 tons) Submerged: 3556 tonnes (3500 tons)
Dimensions:	76.7m x 9.6m x 8.5m (251ft 9in x 31ft 6in x 27ft 10in)
Armament:	Six 533mm (21in) torpedo tubes
Powerplant:	Single shaft, one nuclear PWR, two steam turbines
Range:	Unlimited
Performance:	Surfaced: 18 knots Submerged: 30 knots

Sōryū

The twelve-boat Soryu-class submarines are a series of diesel-electric attack submarines operated by the Japan Maritime Self-Defense Force (JMSDF). They are designed for a variety of missions, including anti-submarine warfare, anti-surface warfare, intelligence gathering, and mine-laying operations. Development began in the early 2000s, with the first, *Soryu* (SS-501), being commissioned in 2009. These submarines are equipped with state-of-the-art technology, including advanced sonar systems, quieting measures, and improved stealth features to enhance their survivability and effectiveness. They also feature an air-independent propulsion (AIP) system, which allows them to operate underwater for longer periods without surfacing. The final two of the class, *Oryu* and *Toryu*, use lithium-ion batteries for greater electric storage capacity and endurance.

Country:	Japan
Launch date:	5 December 2007
Crew:	65
Displacement:	Surfaced: 2900 tonnes (2854 tons) Submerged: 4200 tonnes (4134 tons)
Dimensions:	84m x 9.1m x 8.5m (275ft 7in x 30ft x 27ft 11in)
Armament:	Six 533mm (21in) torpedo tubes; Sub-Harpoon missiles; mines
Powerplant:	Two Kawasaki 12V 25/25 SB-type diesels; 4 Kawasaki-Kockums V4-275R Stirling engines; single screw
Surface range:	11,297km (6100nm) at 6.5 knots
Performance:	Surfaced: 13 knots Submerged: 20 knots

Sturgeon

An enlarged and improved Thresher/Permit design with additional quieting features and electronic systems, the Sturgeon-class SSNs built between 1965 and 1974 were the largest class of nuclear-powered warships until the advent of the Los Angeles class. The Sturgeons were frequently used in the intelligence-gathering role, carrying special equipment and National Security Agency personnel. In 1982 *Cavalla* was converted at Pearl Harbor to have a secondary amphibious assault role by carrying a swimmer delivery vehicle (SDV); *Archerfish*, *Silversides*, *Tunny* and *L. Mendel Rivers* were similarly equipped. *William H. Bates*, *Hawkbill*, *Pintado*, *Richard B. Russell* and others were modified to carry and support the Navy's Deep Submergence and Rescue vehicles. *Parche* (SSN 683) was the last to be decommissioned in July 2005.

Country:	USA
Launch date:	26 February 1966
Crew:	121-141
Displacement:	Surfaced: 4335 tonnes (4266 tons) Submerged: 4854 tonnes (4777 tons)
Dimensions:	89m x 9.65m x 8.9m (292ft 3in x 31ft 8in x 29ft 3in)
Armament:	Four 533mm (21in) torpedo tubes; Tomahawk & Sub Harpoon SSMs
Powerplant:	Single shaft, one nuclear PWR, turbines
Range:	Unlimited
Performance:	Surfaced: 18 knots Submerged: 26 knots

Surcouf

The *Surcouf* was in effect an experimental, one-off boat, described by the French Navy as a 'Corsair submarine'. She was fitted with the largest calibre of guns permitted under the terms of the Washington Treaty, and at the outbreak of war she was the largest, heaviest submarine in the world. She would remain so until the Japanese 400 series entered service in World War II. In June 1940 *Surcouf* escaped from Brest, where she was refitting, and sailed for Plymouth, where she was seized by the Royal Navy (her crew resisting, with casualties on both sides). She was later turned over to the Free Frech Naval Forces, carried out patrols in the Atlantic and took part in the seizure of the islands of St Pierre and Miquelon off Newfoundland. She was lost on 18 February 1942, in collision with an American freighter in the Gulf of Mexico.

Country:	France
Launch date:	18 October 1929
Crew:	118
Displacement	Surfaced: 3302 tonnes (3250 tons)
	Submerged: 4373 tonnes (4304 tons)
Dimensions:	110m x 9.1m x 9.07m (360ft 10in x 29ft 9in x 29ft 9in)
Armament:	Two 203mm (8in) guns, eight 551mm (21.7in)
	and four 400mm (15.75in) torpedo tubes
Powerplant:	Twin screws, diesel/electric motors
Surface range:	18,530km (10,000nm) at 10 knots
Performance	Surfaced: 18 knots
	Submerged: 8.5 knots

Swiftsure

There were six boats in this SSN class, completed between July 1974 and March 1981; *Swiftsure, Sovereign, Superb, Sceptre, Spartan* and *Splendid*. All underwent major refits in the late 1980s or early 1990s, with a full tactical weapons system upgrade. Each was fitted with a PWR 1 Core Z, providing a 12-year life cycle, although refits remained on an eight-year schedule. Two of the class were usually in refit or maintenance. *Splendid* was the first British submarine to carry a warload of Tomahawk cruise missiles, and used them in the 1999 NATO strikes on Serbia. The sonar fit was the Type 2074 (active/passive search and attack), Type 2007 (passive), Type 2046 (towed array), 2019 (intercept and ranging) and Type 2077 (short range classification). As a result of budget cuts *Swiftsure* paid off in 1992, but the others were decommissioned in the 2000s.

Country:	Britain
Launch date:	7 September 1971
Crew:	116
Displacement:	Surfaced: 4471 tonnes (4400 tons)
	Submerged: 4979 tonnes (4900 tons)
Dimensions:	82.9m x 9.8m x 8.5m (272ft x 32ft 4in x 28ft)
Armament:	Five 533mm (21in) torpedo tubes; Tomahawk and Sub Harpoon SSMs
Powerplant:	Single shaft, nuclear PWR, turbines
Surface range:	Unlimited
Performance:	Surfaced: 20 knots
	Submerged: 30+ knots

Swordfish

Just before the outbreak of World War I, the British Admiralty issued a requirement for a submarine capable of achieving 20 knots on the surface. The result was the diesel-powered *Nautilus* which, launched in 1914, produced very disappointing results. In spite of this, the Admiralty pressed ahead with plans for a 20-knot submarine. A 1912 proposal by the engineer Laurenti was re-examined, and developed further by Scott. *Swordfish's* small funnel was lowered electrically and the well covered by a plate. Closing down the funnel took a minute and a half, the heat inside the submarine proving bearable. *Swordfish* was the first submarine to have an experimental emergency telephone buoy fitted. In 1917, after a few months of trials work, *Swordfish* was converted into a surface patrol craft, and was broken up in 1922.

Country:	Britain
Launch date:	18 March 1916
Crew:	25
Displacement:	Surfaced: 947 tonnes (932 tons)
	Submerged: 1123 tonnes (1105 tons)
Dimensions:	70.5m x 7m x 4.5m (231ft 4in x 23ft x 14ft 9in)
Armament:	Two 533mm (21in) and four 457mm (18in) torpedo tubes
Powerplant:	Twin screws, impulse reaction turbines
Surface range:	5556km (3000nm) at 8.5 knots
Performance:	Surfaced: 18 knots
	Submerged: 10 knots

Swordfish

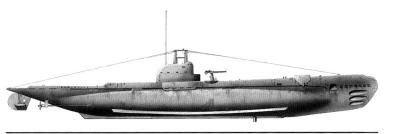

This was the second boat in the first group of S-class submarines. Of the twelve in the group, only four survived World War II. *Swordfish* disappeared without trace off Ushant on or about 10 November 1940, possibly sunk by a mine. Of the other boats, *Seahorse* and *Starfish* were sunk by German minesweepers in the Heligoland Bight; *Shark* was sunk by German minesweepers off Skudesnes, Norway; *Salmon* was mined southwest of Norway; *Snapper* was lost in the Bay of Biscay, cause unknown; *Spearfish* was torpedoed by *U34* off Norway; *Sunfish* was bombed in error by British aircraft while on passage to North Russia (she was to have been assigned to the Soviet Navy as the *B1*); and *Sterlet* was sunk by German trawlers in the Skagerrak. The class leader, *Sturgeon,* served in the Royal Netherlands Navy (1943–45) as the *Zeehond* (Seadog).

Country:	Britain
Launch date:	10 November 1931
Crew:	38
Displacement:	Surfaced: 650 tonnes (640 tons) Submerged: 942 tonnes (927 tons)
Dimensions:	58.8m x 7.3m x 3.2m (193ft x 24ft x 10ft 6in)
Armament:	Six 533mm (21in) torpedo tubes; one 76mm (3in) gun
Powerplant:	Twin shafts, diesel/electric motors
Surface range:	7412km (4000nm) at 10 knots
Performance:	Surfaced: 15 knots Submerged: 10 knots

Tang

The US Navy's Tang class of diesel-electric attack submarines, the equivalents of Russia's Whiskey class, embodied many lessons gleaned from the German Type XXI U-boats. The first four were engined with a new type of radial diesel which caused a lot of problems in service, so they were later re-engined with more conventional motors. The class was conceived virtually on an experimental basis, the US Navy being anxious to discover what improvements could be embodied into a new class of boat in the light of wartime experience and technical advances, but in the event it formed the basis of further postwar development. The six boats in the class were *Trigger* (sold to Italy as the *Livio Piomarto* in 1974), *Wahoo, Trout, Gudgeon* and *Harder* (sold to Italy as the *Romeo Romei* in 1973). *Tang* was sold to Turkey as the *Piri Reis* in 1980.

Country:	USA
Launch date:	19 June 1951
Crew:	83
Displacement:	Surfaced: 1585 tonnes (1560 tons) Submerged: 2296 tonnes (2260 tons)
Dimensions:	82m x 8.3m x 5.2m (269ft 2in x 27ft 2in x 17ft)
Armament:	Eight 533mm (21in) torpedo tubes
Powerplant:	Twin shafts, diesel/electric motors
Surface range:	18,530km (10,000nm) at 10 knots
Performance:	Surfaced: 15.5 knots Submerged: 18.3 knots

Tango

The Tango class of 18 diesel-electric attack submarines was built as an interim measure, filling the gap between the Foxtrot class DE boats and the Victor nuclear-powered SSNs. The first unit was completed at Gorky in 1972, and over the next ten years 17 more units were built in two slightly different versions. The later type were several metres longer than the first in order to accommodate the fire-control systems associated with the tube-launched SS-N-15 anti-submarine missile, the equivalent of the US Navy's SUBROC. The bow sonar installations were the same as those fitted to later Soviet SSNs, while the powerplant was identical to that installed in the later units of the Foxtrot class. Production of the Tango class ceased after the 18th unit was built; the last was decommissioned in 2016, although three are preserved as museum ships.

Country:	Russia
Launch date:	1971
Crew:	60
Displacement:	Surfaced: 3251 tonnes (3200 tons) Submerged: 3962 tonnes (3900 tons)
Dimensions:	92m x 9m x 7m (301ft 10in x 29ft 6in x 24ft)
Armament:	Six 533mm (21in) torpedo tubes
Powerplant:	Twin screws, diesel/electric motors
Surface range:	22,236km (12,000nm) at 10 knots
Performance:	Surfaced: 20 knots Submerged: 16 knots

Terrible

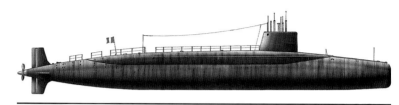

L aid down at Cherbourg Naval Dockyard on 24 June 1967, *Le Terrible* was the second unit of France's Redoutable-class SSBNs, forming the seaborne element of the French *Force de Dissuasion* (Deterrent Force), which ultimately consisted of boats armed with the MSBS (*Mer-Sol Ballistique Strategique*) IRBM. The French term for SSBN is SNLE (*Sous-marin Nucléaire Lance-Engins*). *Le Redoutable* reached IOC in December 1971; *Le Terrible* followed in 1973, *Le Foudroyant* in 1974, *L'Indomptable* in 1976 and *Le Tonnant* in 1979. The upgraded *L'Inflexible* followed in 1985. The first operational launch of an M4 SLBM was made by *Tonnant* on 15 September 1987 in the Atlantic. *Le Redoutable* was withdrawn in 1991; from 1985 the others were all upgraded to match *L'Inflexible*. The latter was the last to be stricken in 2008, replaced by the Triomphant-class SSBNs.

Country:	France
Launch date:	12 December 1969
Crew:	114
Displacement:	Surfaced: 7620 tonnes (7500 tons) Submerged: 9144 tons (9000 tons)
Dimensions:	128m x 10.6m x 10m (420ft x 34ft 10in x 32ft 10in)
Armament:	16 submarine-launched MRBMs
Powerplant:	One nuclear PWR, turbines
Range:	Unlimited
Performance:	Surfaced: 20 knots Submerged: 28 knots

Thames

Designed with a high surface speed for Fleet work, *Thames* and her sisters, *Severn* and *Clyde*, were built by Vickers-Armstrong at Barrow-in-Furness. The outbreak of World War II found *Thames* serving with the 2nd Submarine Flotilla, and in the winter of 1939-40 she joined other boats of the flotilla (*Oberon, Triton, Triumph, Thistle, Triad, Trident* and *Truant)* in operations against enemy shipping off Norway. On 23 July 1940 she was sunk by a mine in the Norwegian Sea. *Severn* and *Clyde* served in home waters until 1941, when they deployed to the Mediterranean. On 20 June 1940 *Clyde* (Lt Cdr Ingram) obtained a torpedo hit on the bows of the German battlecruiser *Gneisenau* off Trondheim. Both boats sank a respectable tonnage of enemy ships. In 1944 they deployed to the Eastern Fleet. *Severn* was scrapped at Bombay and *Clyde* at Durban in 1946.

Country:	Britain
Launch date:	26 January 1932
Crew:	61
Displacement:	surfaced: 1834 tonnes (1805 tons) submerged: 2680 tonnes (2723 tons)
Dimensions:	99.1m x 8.5m x 4.1m (325ft x 28ft x 13ft 6in)
Armament:	Eight 533mm (21in) torpedo tubes; one 100mm (4in) gun
Powerplant:	Two shafts, diesel/electric motors
Surface range:	9265km (5000nm) at 10 knots
Performance:	surfaced: 21.75 knots submerged: 10 knots

Thistle

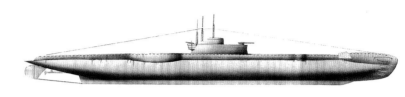

Thistle was one of 22 submarines of the T-class (first group), the first of which, Triton, was launched in October 1937 and the last, Trooper, in March 1942. Most saw active service in the Mediterranean. No fewer than fourteen of this group were lost between September 1939 and October 1943; Thistle herself was sunk off Norway by the U4 on 10 April 1940. Of the remainder, Triton was sunk in the Adriatic by the Italian TB Clio; Thunderbolt (formerly Thetis) was sunk by the corvette Cicogna off Sicily; Tarpon was sunk in the North Sea by the German minesweeper M6; Triumph, Tigris, Triad, Talisman, Tetrarch, Traveller and Trooper were all lost through unknown causes in the Mediterranean; Tempest was sunk by the TB Circe in the Gulf of Taranto; Thorn by the TB Pegaso off Tobruk; and Turbulent by Italian MTBs off Sardinia.

Country:	Britain
Launch date:	25 October 1938
Crew:	59
Displacement:	Surfaced: 1107 tonnes (1090 tons)
	Submerged: 1600 tonnes (1575 tons)
Dimensions:	80.8m x 8m x 4.5m (265ft x 26ft 6in x 14ft 9in)
Armament:	Ten 533mm (21in) torpedo tubes; one 100mm (4in) gun
Powerplant:	Twin screws, diesel/electric motors
Surface range:	7041km (3800nm) at 10 knots
Performance:	Surfaced: 15.25 knots
	Submerged: 9 knots

Thresher/Permit

The first of the SSNs in the US Navy with a deep-diving capability, advanced sonars mounted in the optimum bow position, amidships angled torpedo tubes with the SUBROC ASW missile, and a high degree of machinery-quieting, the Thresher class formed an important part of the US subsurface attack capability for 20 years. The lead boat, USS *Thresher*, was lost with all 129 crew on board, off New England on 10 April 1963, midway through the building period 1960-66. The class was then renamed after *Permit*, the second ship. As a result of the enquiry following the loss of Thresher, the last three of the class were modified during construction with SUBSAFE features. The names of the class were *Permit, Plunger, Barb, Pollack, Haddo, Guardfish, Flasher* and *Haddock* (Pacific Fleet); and *Jack, Tinosa, Dace, Greenling* and *Gato* (Atlantic Fleet).

Country:	USA
Launch date:	9 July 1960 (*Thresher*)
Crew:	134-141
Displacement:	Surfaced: 3810 tonnes (3750 tons)
	Submerged: 4380 tonnes (4311 tons)
Dimensions:	84.9m x 9.6m x 8.8m (278ft 6in x 31ft 8in x 28ft 10in)
Armament:	Four 533mm (21in) torpedo tubes
Powerplant:	Single shaft, one nuclear PWR, steam turbines
Range:	Unlimited
Performance:	Surfaced: 18 knots
	Submerged: 27 knots

Torbay

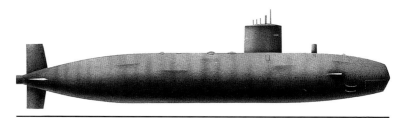

HMS *Torbay* was the fourth boat of the Trafalgar-class SSNs, the lead ship of which was ordered on 7 April 1977 and launched on 1 July 1981. The other vessels in the class were *Turbulent* (launched 1 December 1982), *Tireless* (17 March 1984), *Trenchant* (3 November 1986), *Talent* (15 April 1988) and *Triumph* (16 February 1991). HMS *Trafalgar* was trials submarine for the Spearfish torpedo, which went into full production in 1992 and was first deployed in *Trenchant* early in 1994. The submarines are covered with conformal anechoic coatings to reduce noise. Trafalgars saw active service firing Tomahawks at land targets during the wars in Afghanistan, Iraq and Libya. *Torbay* was decommissioned in 2017; *Triumph*, the last of the class still operational, is due for decommissioning in 2024, replaced by the Astute class SSNs.

Country:	Britain
Launch date:	8 March 1985
Crew:	130
Displacement:	Surfaced: 4877 tonnes (4800 tons) Submerged: 5384 tonnes (5300 tons)
Dimensions:	85.4m x 10m x 8.2m (280ft 2in x 33ft 2in x 27ft)
Armament:	Five 533mm (21in) torpedo tubes; Tomahawk and Sub Harpoon SSMs
Powerplant:	Pump jet, one PWR, turbines
Range:	Unlimited
Performance:	Surfaced: 20 knots Submerged: 32 knots

Trafalgar

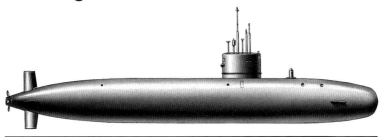

Essentially an improved *Swiftsure*, HMS *Trafalgar* and her sister ships consituted the third generation of British SSNs to be built by Vickers Shipbuilding and Engineering Ltd (VSEL) at Barrow-in-Furness. Improvements included a new reactor system and a pump-jet propulsion system in place of a conventional propeller. *Trafalgar* was the first boat to be fitted with the Type 2020 sonar, and was used as development test platform. The deployment of the Trafalgar class brought the Royal Navy's SSN fleet to 12 boats by the mid-1990s. In 2001, *Trafalgar* was involved in Operation Veritas, firing Tomahawk missiles at Al-Qaeda and Taliban forces in Afghanistan. She ran aground near the Isle of Skye in 2002, causing £5m of damage during a training exercise. She was decommissioned in 2009.

Country:	Britain
Launch date:	1 July 1981
Crew:	130
Displacement:	Surfaced: 4877 tonnes (4800 tons)
	Submerged: 5384 tonnes (5300 tons)
Dimensions:	85.4m x 10m x 8.2m (280ft 2in x 33ft 2in x 27ft)
Armament:	Five 533mm (21in) torpedo tubes; Tomahawk
	and Sub Harpoon SSMs
Powerplant:	Pump jet, one PWR, turbines
Range:	Unlimited
Performance:	Surfaced: 20 knots
	Submerged: 32 knots

Le Triomphant

*L*e *Triomphant* is the lead ship of the latest class of French nuclear-powered ballistic missile submarines, all built by DCN Cherbourg. She was ordered on 10 March 1986 and commissioned on 21 March 1997. The second, *Le Teméraire,* was launched on 21 January 1998, while the third, *Le Vigilant,* was launched on 19 September 2003. A fourth vessel, *Le Terrible,* was launched on 21 March 2008, and commissioned on 20 September 2010. *Le Triomphant* began sea trials in April 1994, and her first sea cruise took place from 16 July to 22 August 1995. The first submerged launch of an M45 SLBM (range 8000km/4300nm, 6 MRVs) was made on 14 February 1995. This was replaced with the upgraded M51 SLBM that entered service in 2010. In February 2009, *Le Triomphant* damaged her bow after colliding with the British Royal Navy's HMS *Vanguard*.

Country:	France
Launch date:	13 July 1993
Crew:	111
Displacement:	Surfaced: 12,842 tonnes (12640 tons)
	Submerged: 14,335 tonnes (14,565 tons)
Dimensions:	138m x 17m x 12.5m (453ft x 77ft 9in x 41ft)
Armament:	Sixteen M45/TN75 SLBM; four 533mm (21in) torpedo tubes
Powerplant:	Pump jet, one nuclear PWR, two diesels
Range:	Unlimited
Performance:	Surfaced: 20 knots
	Submerged: 25 knots

Triton

U SS *Triton* was designed and built for use as a radar picket submarine to operate in conjunction with surface carrier task forces, submerging only when in danger of enemy attack. For this purpose she was fitted with an elaborate combat information centre and a large radar antenna that retracted into her sail. At that time she was the longest submarine ever built, and was exceeded in displacement only by the later Polaris SSBNs. In 1960 *Triton* circumnavigated the globe entirely underwater, except for one instance when her sail structure broke surface to allow a sick submariner to be taken off near the Falkland Islands. The 66,749km (36,022nm) cruise took 83 days and was made at an average speed of 18 knots. On 1 March 1961 *Triton* was reclassified as an attack submarine. She was decommissioned on 3 May 1969.

Country:	USA
Launch date:	19 August 1958
Crew:	172
Displacement:	Surfaced: 6035 tonnes (5940 tons) Submerged: 7905 tonnes (7780 tons)
Dimensions:	136.3m x 11.3m x 7.3m (447ft 6in x 37ft x 24ft)
Armament:	Six 533mm (21in) torpedo tubes
Powerplant:	Two shafts, one nuclear PWR, turbines
Range:	Unlimited
Performance:	Surfaced: 27 knots Submerged: 20 knots

Turtle

Although not a true submersible, all of *Turtle* was under the surface when she was in action except for a tiny conning tower fitted with glass ports so that the occupant could find his way to the target. She was built by David Bushnell who, during the American War of Independence, had carried out experiments with floating tide-borne mines. He subsequently decided to built a manned semi-submersible craft that could transport an explosive charge to the hull of an enemy ship. The charge was carried outside the hull, and was attached to the target by an auger that was drilled into the hull of the enemy vessel. A clockwork timing device gave *Turtle* time to escape. In September 1776 an American soldier, Ezra Lee, tried to attach an explosive charge to HMS *Eagle* in the Hudson River, but the auger broke. It was the first-ever underwater war mission.

Country:	USA
Launch date:	1776
Crew:	1
Displacement:	Surfaced: 2 tonnes/2 tons Submerged: 2 tonnes/2 tons
Dimensions:	1.8m x 1.3m (6ft x 4ft 6in)
Armament:	One 68kg (150lb) detachable explosive charge
Powerplant:	Single screw, hand-cranked
Surface range:	Not known
Performance:	Surfaced: Not known Submerged: Not known

Type 039

Given the reporting name of Song by NATO, the Type 039 – an ocean-going multi-role SSK – is the first submarine to have been completely designed, built and fitted out in China. Thirteen vessels were built between 1999 and 2006, in three variants. The first, No. 320, was laid down in 1991. It was launched in May 1994, but was not commissioned until June 1999 due to numerous design and performance problems. The second in the class, the first improved Type 039G, No. 321, was launched in November 1999 and commissioned in April 2001, with a total of five of the variant built. The first of the further improved Type 039G1 design was commissioned in 2003. All the variants use German diesels as their power source. Further development of the Song class led to the Type 039A/041 Yuan class, the first Chinese design with air-independent propulsion.

Country:	China
Launch date:	May 1994
Crew:	60
Displacement:	Surfaced: 1700 tonnes (1673 tons) Submerged: 2286 tonnes (2250 tons)
Dimensions:	74.9m x 8.4m x 5.3m (245ft 8in x 27ft 6in x 17ft 4in)
Armament:	Six 533mm (21in) torpedo tubes, 18 torpedoes/missiles or 24–36 mines
Powerplant:	Four MTU 16V 396S E84 diesels, 24,360hp (18,148kW); single screw
Surface range:	N/A
Performance:	Surfaced: 15 knots Submerged: 22 knots

Type 094

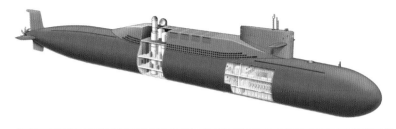

The Type 094 is the second class of nuclear-powered ballistic missile submarines (SSBNs) developed and operated by the People's Liberation Army Navy (PLAN) of China. Each Type 094 submarine can carry up to 12 JL-2 or JL-3 submarine-launched ballistic missiles (SLBMs). The JL-3 missiles are believed to be equipped with three multiple independently targetable reentry vehicles (MIRV) warheads and have a range capable of reaching targets across the United States. Four boats, *Changzeng 11*, *12*, *13* and *14*, entered service between 2007 and 2015 before a new variant with an improved sail was introduced as the Type 094A, two of which, *Changzeng 20* and *21*, are currently active. The class is due to be augmented in service sometime in the late 2020s by the new Type 096 class now under development.

Country:	China
Launch date:	28 July 2004
Crew:	140
Displacement:	Surfaced: 8128 tonnes (8000 tons)
	Submerged: 11,176 tonnes (11,000 tons)
Dimensions:	135m x 12.5m x 8m (442ft 11in x 41ft 26ft 3in)
Armament:	Twelve SLBMs; six 533mm (21in) torpedo tubes
Powerplant:	One nuclear PWR, single shaft
Range:	Unlimited
Performance:	Surfaced: N/A
	Submerged: 20 knots

Type 212A

The Type 212A is a class of diesel-electric attack submarines developed by HDW for the German and Italian navies. Featuring an air-independent propulsion (AIP) system, the Type 212A submarines can operate submerged for extended periods without the need to surface or snorkel, greatly enhancing their operational endurance and stealthiness. This AIP system, utilizing hydrogen fuel cells, provides a significant advantage over traditional diesel-electric submarines, enabling longer submerged patrols and reducing the risk of detection. The German Navy operates six Type 212A submarines – *U31* to *U36*, while the Italian Navy operates four as the Todaro class – *Salvatore Todaro*, *Scirè*, *Pietro Venuti* and *Romeo Romei* – with at least another three on order. A significantly enlarged variant, the Type 212CD, has been ordered by Norway and Germany.

Country:	Germany
Launch date:	20 March 2002
Crew:	27
Displacement:	Surfaced: 1450 tonnes (1430 tons)
	Submerged: 1830 tonnes (1800 tons)
Dimensions:	56m x 7m x 6m (183ft x 23ft x 19ft 8in)
Armament:	Six 533mm (21in) tubes; 24 torpedoes
Powerplant:	One diesel generator, one electric motor, single shaft
Surface range:	14,816 (9206nm) at 8 knots
Performance:	Surfaced: 12 knots
	Submerged: 20 knots

Typhoon

Capable of hitting strategic targets anywhere in the world with its Makayev SS-N-20 Sturgeon three-stage solid fuel missiles, each of which has 10 200kT nuclear warheads and a range of 8300km (4500nm), Typhoon (Project 941 *Akula*) was the largest class of submarine ever built. The launch tubes were positioned forward in the bow section, leaving space abaft the fin for two nuclear reactors. The fin could break through ice up to 3m (9ft 10in) thick, and diving depth was in the order of 300m (1000ft). Six Typhoons were commissioned between 1981 and 1989; their designations, in order, were TK208, TK202, TK12, TK13, TK17 and TK20. TK17 was damaged by fire in a missile loading accident in 1992 but was subsequently repaired. All but the first were decommissioned by 2013; *Dmitriy Donskoy* (TK208) survived until 2023 as a weapons test platform.

Country:	Russia
Launch date:	23 September 1980
Crew:	175
Displacement:	Surfaced: 18,797 tonnes (18,500 tons)
	Submerged: 26,925 tonnes (26,500 tons)
Dimensions:	171.5m x 24.6m x 13m (562ft 7in x 80ft 7in x 42ft 6in)
Armament:	Twenty SLBMs; four 630mm (25in)
	and two 533mm (21in) torpedo tubes
Powerplant:	Two shafts, two nuclear PWR, turbines
Range:	Unlimited
Performance:	Surfaced: 12 knots
	Submerged: 25 knots

U1

Strangely enough, the German Naval Staff at the turn of the century failed to appreciate the potential of the submarine, and the first submarines built in Germany were three Karp-class vessels ordered by the Imperial Russian Navy in 1904. Germany's first practical submarine, *U1*, was not completed until 1906. She was, however, one of the most successful and reliable of the period. Her two kerosene engines developed 400hp, as did her electric motors. She had an underwater range of 80km (43nm). Commissioned in December 1906, she was used for experimental and training purposes. In February 1919 she was stricken, sold, and refitted as a museum exhibit by her builders, Germaniawerft of Kiel. She was damaged by bombing in World War II, but subsequently restored. Double hulls and twin screws were used from the first in German U-boats.

Country:	Germany
Launch date:	4 August 1906
Crew:	22
Displacement:	Surfaced: 241 tonnes (238 tons) Submerged: 287 tonnes (283 tons)
Dimensions:	42.4m x 3.8m x 3.2m (139ft x 12ft 6in x 10ft 6in)
Armament:	One 450mm (17.7in) torpedo tube
Powerplant:	Twin screws, heavy oil (kerosene)/electric motors
Surface range:	2850km (1536nm) at 10 knots
Performance:	Surfaced: 10.8 knots Submerged: 8.7 knots

U2

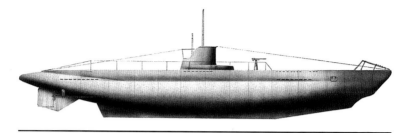

Under the terms of the Versailles Treaty, Germany was forbidden to build submarines, but during the 1920s she set up clandestine design teams in Spain, Holland and Russia. The first boat was built for Finland in 1927 and this was the prototype for *U2*, one of the first Type II submarines intended for coastal service. The diesel engines developed 350hp, and the electric motors developed 180hp. The early Type IIs were all used for training, and some of the most talented and successful of Germany's submarine commanders in World War II learned their trade in them. In March 1940 *U2* joined other German U-boats in an unsuccessful 'hunter-killer' operation against British and French submarines in the North Sea. On 8 April 1944 *U2* was lost in a collision west of Pillau during a training sortie in the Baltic.

Country:	Germany
Launch date:	July 1935
Crew:	25
Displacement:	Surfaced: 254 tonnes (250 tons) Submerged: 302 tonnes (298 tons)
Dimensions:	40.9m x 4.1m x 3.8m (133ft 2in x 13ft 5in x 12ft 6in)
Armament:	Three 533mm (21in) torpedo tubes, one 20mm (0.8in) AA gun
Powerplant:	Twin screws, diesel/electric motors
Surface range:	1688km (912nm) at 10 knots
Performance:	Surfaced: 13 knots Submerged: 7 knots

U3

As early as 1922 the Germans set up a submarine design office at Den Haag (The Hague) in the Netherlands under cover of a Dutch firm. It was under the guise of constructing submarines for foreign navies that the German designers and constructors – who had remained in close touch since the end of World War I – set about producing craft which would actually serve as prototypes for a reborn German Navy. The last of five boats for Finland, the *Vessiko*, was built by the German firm Chrichton-Vulcan AB at Turku on the southwestern tip of Finland and given the designation Submarine 707, although she was actually the prototype for the Type IIA U-boat. *U3* was a Type IIA. Because of her limited range, she was used mainly for training. She was paid off at Gdynia in July 1944 and cannibalized for spare parts early in 1945.

Country:	Germany
Launch date:	1936
Crew:	25
Displacement:	Surfaced: 254 tonnes (250 tons)
	Submerged: 302 tonnes (298 tons)
Dimensions:	40.9m x 4.1m x 3.8m (133ft 2in x 13ft 5in x 12ft 6in)
Armament:	Three 533mm (21in) torpedo tubes, one 20mm (0.8in) AA gun
Powerplant:	Twin screws, diesel/electric motors
Surface range:	1688km (912nm) at 10 knots
Performance:	Surfaced: 13 knots
	Submerged: 7 knots

U12

U^{12} was the last of the German Federal Republic's Type 205 coastal submarines, the fourth such class to become operational since the FDR rearmed as part of NATO in the 1950s. The first was the *Hai* (Shark) class, comprising that boat and the *Hecht* (Pike), both of which were reconstructed World War II Type XXIII U-boats. These were followed by the Types 201 and 202. With the exception of two boats all were to have been of the Type 201 model, but there were severe hull corrosion problems, the non-magnetic material used proving quite unsatisfactory. As an interim measure the hulls of *U4* to *U8* were covered in tin, suffering severe operational limitations as a consequence, and construction of *U9* to *U12* was suspended until a special non-magnetic steel could be developed.

Country:	Germany
Launch date:	10 September 1968
Crew:	21
Displacement:	Surfaced: 425 tonnes (419 tons)
	Submerged: 457 tonnes (450 tons)
Dimensions:	43.9m x 4.6m x 4.3m (144ft x 15ft x 14ft)
Armament:	Eight 533mm (21in) torpedo tubes
Powerplant:	Single screw, diesel/electric motors
Surface range:	7041km (3800nm) at 10 knots
Performance:	Surfaced: 10 knots
	Submerged: 17.5 knots

U21

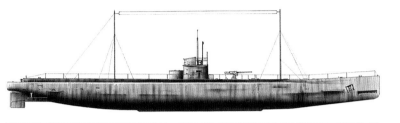

The *U21* was one of a class of four U-boats, built at Danzig and completed in 1913. Although the Germans got away to a slow start in their submarine construction programme before World War I, the vessels were well engineered and used double hulls and twin screws from the start. German engineers refused to employ petrol engines in the early boats, preferring to use smellier but safer kerosene fuel. In 1908 suitable diesel engines were designed, and these were installed in the *U19* class (to which *U21* belonged) and used exclusively thereafter. Of the four boats in the class, *U19* and *U22* surrendered in November 1918 and were scrapped at Blyth, Northumberland; *U20* was scuttled after being stranded on the Danish coast in 1916, and broken up in 1925; and *U21* foundered in the North Sea on 22 February 1919 as she was sailing to surrender.

Country:	Germany
Launch date:	8 February 1913
Crew:	35
Displacement:	Surfaced: 660 tonnes (650 tons) Submerged: 850 tonnes (837 tons)
Dimensions:	64.2m x 6.1m x 3.5m (210ft 6in x 20ft x 11ft 9in)
Armament:	Four 508mm (20in) torpedo tubes; one 86mm (3.4in) gun
Powerplant:	Two shafts, diesel/electric motors
Surface range:	9265km (5500nm) at 10 knots
Performance:	Surfaced: 15.4 knots Submerged: 9.5 knots

U28

Studies for a replacement submarine to follow the Type 205 class were initiated in 1962. The result was the new Type 206 class which, built of high-tensile non-magnetic steel, was to be used for coastal operations and had to conform with treaty limitations on the maximum tonnage allowed to West Germany. New safety devices were fitted, and the armament fit permitted the carriage of wire-guided torpedoes. Construction planning took place in 1966–68, and the first orders (for an eventual total of 18 units) were placed in 1969. By 1975 all the vessels (*U13* to *U30*) were in service. The submarines were later fitted with two external containers for up to 24 ground mines. Four were sold to Columbia, two for spares, and two entered service in 2015: ARC *Intrépido* (ex *U23*) and ARC *Indomable* (ex *U24*). The last German Navy boat was decommissioned in 2011.

Country:	Germany
Launch date:	22 January 1974
Crew:	21
Displacement:	Surfaced: 457 tonnes (450 tons) Submerged: 508 tonnes (500 tons)
Dimensions:	48.6m x 4.6m x 4.5m (159 ft 5in x 15ft 2in x 14ft 10in)
Armament:	Eight 533mm (21in) torpedo tubes
Powerplant:	Single shaft, diesel/electric motors
Surface range:	7041km (3800nm) at 10 knots
Performance:	Surfaced: 10 knots Submerged: 17 knots

U32

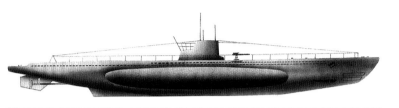

In September 1939 the German Navy had only 56 submarines in commission, and of these only 22 were ocean-going craft, suitable for service in the Atlantic. They were Type VIIs, the class to which *U32* belonged. With a conning tower only 5.2m (17ft) above the waterline they were hard to detect even in daylight, and under night conditions they were practically invisible. They could dive in less than half a minute; they could go to a depth of 100m (328ft) without strain and 200m (656ft) if hard pressed. They could maintain a submerged speed of 7.6 knots for two hours, or two knots for 130 hours. In fact, their depth and endurance performance at high speed was twice as good as that of any other submarine. *U32* was sunk in the North Atlantic on 30 October 1940 by the RN destroyers *Harvester* and *Highlander*.

Country:	Germany
Launch date:	1937
Crew:	44
Displacement:	Surfaced: 636 tonnes (626 tons)
	Submerged: 757 tonnes (745 tons)
Dimensions:	64.5m x 5.8m x 4.4m (211ft 6in x 19ft 3in x 14ft 6in)
Armament:	Five 533mm (21in) torpedo tubes; one 88mm (3.5in) gun;
	one 20mm AA gun
Powerplant:	Two screws, diesel/electric motors
Surface range:	6916km (3732nm) at 12 knots
Performance:	Surfaced: 16 knots
	Submerged: 8 knots

U47

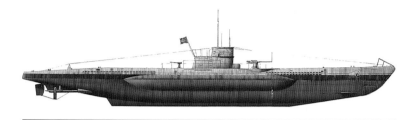

The Type VIIB U-boat was a slightly enlarged version of the Type VIIA, with a greater range and slightly higher surface speed. The most famous boat of this class was undoubtedly the *U47*, commanded by Lt Cdr Gunther Prien, who on the night of 13-14 October 1939 penetrated the defences of Scapa Flow and sank the 27,940-tonne (27,500-ton) battleship *Royal Oak*, a veteran of World War I, with three torpedo hits. The attack, in which 833 lives were lost, was carried out with great coolness, skill and daring, and came as a great shock to Britain. Prien returned home to a hero's welcome. He had already sunk three small merchant ships on the first day of the war, and went on to sink 27 more before *U47* was sunk in the North Atlantic on 7 March 1941 by the RN corvettes *Arbutus* and *Carmellia*.

Country:	Germany
Launch date:	1938
Crew:	44
Displacement:	Surfaced: 765 tonnes (753 tons) Submerged: 871 tonnes (857 tons)
Dimensions:	66.5m x 6.2m x 4.7m (218ft x 20ft 3in x 15ft 6in)
Armament:	Five 533mm (21in) torpedo tubes; one 88mm (3.5in) gun; one 20mm AA gun
Powerplant:	Two-shaft diesel/electric motors
Surface range:	10,454km (5642nm) at 12 knots
Performance:	Surfaced: 17.2 knots Submerged: 8 knots

U106

The Type IXB U-boats, of which *U106* was one, were improvements of the ocean-going Type IXAs with increased radius. Some Type IXBs were modified for service in the Far East, their range being increased to 16,100km (8700nm) at 12 knots. Their hunting ground was the Indian Ocean, using bases in Japanese-occupied Malaya and Singapore for replenishment. In March 1941 the *U106* (Lt Cdr Oesten), having already sunk several merchant ships on her Atlantic patrols, torpedoed the British battleship *Malaya*, which was escorting a convoy. The battleship was repaired in New York, but was effectively out of the war. Under Capt Rasch, *U106* went on to sink many more merchantmen in the Atlantic before being destroyed by air attack off Cape Ortegal, Biscay, on 2 August 1943. Their high surface speed made the Type IXs very effective in surface night attacks.

Country:	Germany
Launch date:	1939
Crew:	48
Displacement:	Surfaced: 1068 tonnes (1051 tons) Submerged: 2183 tonnes (1178 tons)
Dimensions:	76.5m x 6.8m x 4.6m (251ft x 22ft 3in x 15ft)
Armament:	Six 533mm (21in) torpedo tubes; one 102mm (4.1in) gun, one 20mm AA gun
Powerplant:	Twin shafts, diesel/electric motors
Surface range:	13,993km (7552nm) at 10 knots
Performance:	Surfaced: 18.2 knots Submerged: 7.2 knots

U112

During World War I, as their primary aim was to sink merchant ships, German submarines were fitted with progressively heavier armament until eventually, boats of new construction carried two 150mm (5.9in) guns and older boats were modified to bring them up to a similar standard. These weapons could outrange any gun mounted for defensive purposes in merchant vessels, so that U-boats could destroy their victims by gunfire in surface actions, reserving their torpedoes for more dangerous, high-value warship targets. A special class of long-range boat, the submarine cruiser, was designed, and in World War II this concept was resurrected in the Type XI class of U-boat. Designated *U112* to *U115*, the boats were to have carried a spotter aircraft to extend their radius of observation. However, the class never got beyond the project stage.

Country:	Germany
Launch date:	Projected only
Crew:	110
Displacement:	Surfaced: 3190 tonnes (3140 tons)
	Submerged: 3688 tonnes (3630 tons)
Dimensions:	115m x 9.5m x 6m (377ft x 31ft x 20ft)
Armament:	Eight 533mm (21in) torpedo tubes; four 127mm (5in) guns;
	two 30mm and two 20mm AA guns
Powerplant:	Two shafts, diesel/electric motors
Surface range:	25,266km (13,635nm) at 12 knots
Performance:	Surfaced: 23 knots
	Submerged: 7 knots

U139

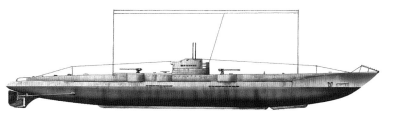

In 1917 the Germans converted two U151-class submarines, *U151* and *U155*, as long-range cargo-carrying vessels. One, the *Deutschland,* made two commercial runs to the United States before America's entry into the war brought an end to the venture; being converted back to naval use along with her sister vessel, *Oldenburg.* The boats were reclassified as submarine cruisers and more were laid down, including *U139, U140* and *U141*. The first two were among the very few German submarines to be given names, probably because of their intended role as surface combatants for the most part; the *U139* became the *Kapitanleutnant Schweiger,* and the *U140 Kapitanleutnant Weddingen*. After the war U139 was allocated to France as the *Halbronn,* U140 was sunk as a gunnery target by an American destroyer, and *U141* was scrapped in 1923.

Country:	Germany
Launch date:	3 December 1917
Crew:	62
Displacement:	Surfaced: 1961 tonnes (1930 tons)
	Submerged: 2523 tonnes (2483 tons)
Dimensions:	94.8m x 9m x 5.2m (311ft x 29ft 9in x 17ft 3in)
Armament:	Six 508mm (20in) torpedo tubes; two 150mm (5.9in) guns
Powerplant:	Twin shafts, diesel/electric motors
Surface range:	23,390km (12,630nm) at 8 knots
Performance:	Surfaced: 15.8 knots
	Submerged: 7.6 knots

U140

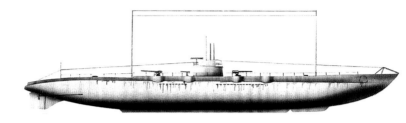

The *U140* mentioned in the previous entry, was one of three 'submarine cruisers' of the U139 class, two of which were named. The Germans also took the unusual step of allocating names to the boats of two succeeding classes, *U145* to *U147* and *U148* to *U150*. These boats, also classed as submarine cruisers, were named *Kapitanleutnant Wegener* (U145), *Oberleutnant-zur-Zee Saltzwedel* (U146), *Kapitanleutnant Hansen* (U147), *Oberleutnant-zur-Zee Pustkuchen* (U148), *Kapitanleutnant Freiherr von Berkheim* (U149) and *Kapitanleutnant Schneider* (U150). The first three boats were laid down at the A.G. Vulcan yard, Hamburg, the others at A.G. Weser, Bremen. All six boats were broken up before completion; *U145-U147* were actually launched in June-September 1918, but by then the war was nearly over.

Country:	Germany
Launch date:	4 November 1917
Crew:	62
Displacement:	Surfaced: 1961 tonnes (1930 tons) Submerged: 2523 tonnes (2483 tons)
Dimensions:	94.8m x 9m x 5.2m (311ft x 29ft 9in x 17ft 3in)
Armament:	Six 508mm (20in) torpedo tubes; two 150mm (5.9in) guns
Powerplant:	Twin shafts, diesel/electric motors
Surface range:	32,873km (17,750nm) at 8 knots
Performance:	Surfaced: 15.8 knots Submerged: 7.6 knots

U151

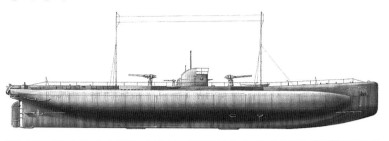

Before America's entry into the war in 1917 the Germans were quick to recognize the potential of large, cargo-carrying submarines as a means of beating the blockade imposed on Germany's ports by the Royal Navy. Two U151-class submarines, the *U151* and *U155*, were converted for mercantile use and named *Oldenburg* and *Deutschland* respectively. After America's entry into the war the two boats were converted back to naval use as heavily-armed submarine cruisers, forming two of a class of seven (*U151-U157*). On 24 November 1918 *U151* was surrendered to France and was sunk as a target vessel off Cherbourg on 7 June 1921. *U155*, formerly *Deutschland*, was scrapped at Morecambe, England, in 1922, while a third merchant conversion, *Bremen*, was lost on her first voyage in 1917, possibly mined off the Orkneys.

Country:	Germany
Launch date:	4 April 1917
Crew:	56
Displacement:	Surfaced: 1536 tonnes (1512 tons) Submerged: 1905 tonnes (1875 tons)
Dimensions:	65m x 8.9m x 5.3m (213ft 3in x 29ft 2in x 17ft 5in)
Armament:	Two 509mm (20in) torpedo tubes; two 150mm (5.9in) and two 86mm (3.4in) guns
Powerplant:	Twin screw diesel engines, electric motors
Range:	20,909km (11,284nm) at 10 knots
Performance:	Surfaced: 12.4 knots Submerged: 5.2 knots

U160

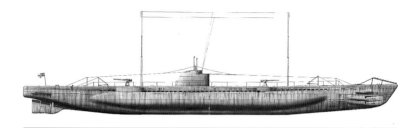

The *U160* was leader of a class of 13 fast U-boats laid down in the last months of the war. She was built by Bremer Vulcan at Kiel. Five of these boats, *U168* to *U172*, were scrapped before they were completed. Of the remainder, *U160* was surrendered to France and scrapped at Cherbourg in 1922; *U161* was stranded en route to the breakers; *U162* was also surrendered to the French, serving as the *Pierre Marast* before being scrapped in 1937; *U163* was handed over to Italy and scrapped in 1919; U164 was scrapped at Swansea in 1922; *U165* was sunk by accident in the Weser; *U166* was completed after the Armistice and handed over to France, serving as the *Jean Roulier* before going to the breaker's yard in 1935; and *U167* was scrapped in 1921. Because of their high speed, boats of this class usually attacked on the surface.

Country:	Germany
Launch date:	27 February 1918
Crew:	39
Displacement:	Surfaced: 834 tonnes (821 tons)
	Submerged: 1016 tonnes (1000 tons)
Dimensions:	71.8m x 6.2m x 4.1m (235ft 6in x 20ft 6in x 13ft 6in)
Armament:	Six 509mm (20in) torpedo tubes; two 104mm (4.1in) guns
Powerplant:	Two shafts, diesel/electric motors
Surface range:	15,372km (8300nm) at 8 knots
Performance:	Surfaced: 16.2 knots
	Submerged: 8.2 knots

U1081

The *U1081* was leader of a planned class of ten Type XVIIG coastal U-boats powered by the revolutionary Walter geared turbine, but still fitted with diesel/electric drive to extend their radius of action. The Type XVIIG class were generally similar to the Type XVIIB, although about 1.5m (5ft) shorter. A further class of experimental boats, the Type XVIIK, was planned for the purpose of testing the closed-cycle diesel engine as an alternative to the Walter turbine, but like the XVIIG it never advanced beyond project status. Three XVIIBs were built, all being scuttled in May 1945; one of them, the *U1407*, was salved, repaired and allocated to the Royal Navy under the name *Meteorite*. She was used to make exhaustive tests of the Walter propulsion system and was scrapped in 1950. Had they become available, the Type XV11s would have been formidable opponents.

Country:	Germany
Launch date:	Project cancelled (1945)
Crew:	19
Displacement:	Surfaced: 319 tonnes (314 tons)
	Submerged: 363 tonnes (357 tons)
Dimensions:	40.5m x 3.3m x 4.3m (129ft 6in x 11ft x 14ft)
Armament:	Two 533mm (21in) torpedo tubes
Powerplant:	Single-shaft geared turbine; diesel/electric motors
Surface range:	Not known
Performance:	Surfaced: 23 knots (estimated)
	Submerged: 8.5 knots

U2326

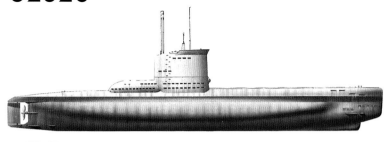

In the latter months of the war Germany launched a massive submarine construction programme, the aim of which was to get two types of submarine – the Type XXI and Type XXIII – into service as quickly as possible. Both were fitted with diesel/electric motors plus electric 'creeping' motors that made them extremely hard to detect. The *U2326* was a Type XXIII U-boat, one of 57 that were either at sea, in various stages of construction or projected at the end of the war in Europe. The building programme was severely disrupted by Allied bombing, and only a few Type XXIIIs were operational in the final weeks of the war. After the surrender the *U2326* went to Britain, where she was used for experimental work as the N25. She was handed over to France in 1946, and lost in an accident off Toulon in December that year.

Country:	Germany
Launch date:	Not known
Crew:	14
Displacement:	Surfaced: 236 tonnes (232 tons)
	Submerged: 260 tonnes (256 tons)
Dimensions:	34m x 2.9m x 3.7m (112ft 9in x 9ft 9in x 12ft 3in)
Armament:	Two 533mm (21in) torpedo tubes
Powerplant:	Single-shaft diesel/electric motors; silent creeping electric motor
Surface range:	2171km (1172nm) at 7 knots
Performance:	Surfaced: 9.75 knots
	Submerged: 12.5 knots

U2501

U *2501*, the first of the Type XXI ocean-going U-boats, was a milestone in the development of the submarine, a stop on the evolutionary road that led to the nuclear-powered vessels of today. She was a double-hulled vessel, with high submerged speed plus the ability to run silently at 3.5 knots on her 'creeper' electric motors. The outer hull was built of light plating to aid streamlining; the inner hull was 28-37mm (1.1–1.5in) thick carbon steel plating. She had new, super-light batteries, and could maintain a submerged speed of 16 knots for one hour; at four knots, she could remain submerged for three days on a single charge. Some 55 Type XXIs were in service when Germany surrendered in 1945, but a great many were destroyed by bombing during construction – fortunately for the Allies, for they were very dangerous war machines.

Country:	Germany
Launch date:	1944
Crew:	57
Displacement:	Surfaced: 1647 tonnes (1621 tons) Submerged: 2100 tonnes (2067 tons)
Dimensions:	77m x 8m x 6.2m (251ft 8in x 26ft 3in x 20ft 4in)
Armament:	Six 533mm (21in) torpedo tubes; four 30mm (1.2in) AA guns
Powerplant:	Twin screws, diesel/electric motors, silent creeping motors
Surface range:	17934km (9678nm) at 10 knots
Performance:	Surfaced: 15.5 knots Submerged: 16 knots

U2511

As soon as they were commissioned, the ocean-going Type XXI and the coastal Type XXIII U-boats were deployed to Norway, where it was anticipated that the German armed forces would make a last-ditch defence. The German surrender on 8 May 1945 found the Type XXI *U2511*, the first such boat to become operational in March 1945, in the Norwegian harbour of Bergen, having just returned from a patrol in which a British cruiser was sighted on 4 May. The *U2511*'s captain, advised that the German surrender was imminent, elected to carry out only a dummy attack on the British warship, so saving its crew and probably his own. The Royal Navy knew about the Type XXIs, the deployment of which caused great concern, but they could not be brought into service fast enough before Germany's final collapse.

Country:	Germany
Launch date:	Late 1944
Crew:	57
Displacement:	Surfaced: 1647 tonnes (1621 tons)
	Submerged: 2100 tonnes (2067 tons)
Dimensions:	77m x 8m x 6.2m (251ft 8in x 26ft 3in x 20ft 4in)
Armament:	Six 533mm (21in) torpedo tubes; four 30mm (1.2in) AA guns
Powerplant:	Twin screws, diesel/electric motors, silent creeping motors
Surface range:	17,934km (9678nm) at 10 knots
Performance:	Surfaced: 15.5 knots
	Submerged: 16 knots

U class

The U-class submarines, 51 boats in all, were built on two groups by Vickers-Armstrong, either at Barrow-in-Furness or on the Tyne. All boats served in home waters and the Mediterranean during World War II, except for five boats which went to the East Indies for ASW training and four which were loaned to the RCN for the same purpose. The boats, which were small and manoeuvrable, proved very effective in the confined waters of the Mediterranean and the Aegean, but war losses were high. *Undine, Unity, Umpire* and *Uredd* (RNN) were lost in home waters; *Unbeaten* in the Bay of Biscay; and *Undaunted, Union, Unique, Upholder, Urge, Usk, Usurper* and *Utmost* in the Mediterranean, together with the unnamed U-class boats *P32, P33, P38* and *P48*. Another boat, *Untamed*, was lost on trials.

Country:	Britain
Launch date:	5 October 1937 (HMS *Undine*, class leader)
Crew:	31
Displacement:	Surfaced: 554 tonnes (545 tons) Submerged: 752 tonnes (740 tons)
Dimensions:	54.9m x 4.8m x 3.8m (180ft x 16ft x 12ft 9in)
Armament:	Four 533mm (21in) torpedo tubes; one 76mm (3in) gun
Powerplant:	Twin screws, diesel/electric motors
Surface range:	7041km (3800nm) at 10 knots
Performance:	Surfaced: 11.2 knot Submerged: 10 knots

UB4

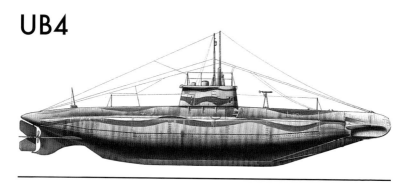

In 1914 the Germans began construction of a new series of U-boats, the small UB coastal class. Upon completion the majority of these were sent by rail in sections to Antwerp in Belgium, which was in German hands, or Pola, the Austro-Hungarian port on the Adriatic, where they were assembled and made ready for deployment. There were no fewer than 25 classes of UB boats, the designs growing rapidly larger as the war progressed. There were eight boats in *UB4*'s class, of which *UB1* was wrecked in the Adriatic and scrapped, *UB2* was stricken in 1919, *UB3* was missing in the Aegean Sea, *UB4* was sunk in the North Sea by the RN armed trawler *Inverlyon* in August 1915, *UB5* was stricken in 1919, *UB6* was surrendered to France, *UB7* was sunk in the Black Sea and *UB8* was handed over to France.

Country:	Germany
Launch date:	April 1915
Crew:	14
Displacement:	Surfaced: 129 tonnes (127 tons) Submerged: 144 tonnes (142 tons)
Dimensions:	28m x 2.9m x 3m (92ft 3in x 9ft 9in x 10ft)
Armament:	Two 457mm (18in) torpedo tubes
Powerplant:	Single screw, diesel/electric motors
Surface range:	2778km (1599nm) at 5 knots
Performance:	Surfaced: 6.5 knots Submerged: 5.5 knots

UC74

UC74 was one of a class of six minelaying submarines, all launched in October-December 1917 and built by A.G. Vulcan, Hamburg. The boats were fitted with six vertical mine tubes. UC74 served briefly with the Austrian Navy as the U93, but with a German crew. At the end of World War I she was interned at Barcelona, having been forced to put in there through lack of fuel, and was surrendered to France in 1919. She was scrapped at Toulon in 1921. Of the other boats, UC75 was sunk in the North Sea by the destroyer HMS *Fairy*; UC76 was accidentally lost off Heligoland when her mines exploded, salvaged and recommissioned, and eventually interned at Karlskrona, Sweden; UC77 and UC78 were sunk by RN drifters in the Dover Straits, while UC79 was mined and sunk in the same area.

Country:	Germany
Launch date:	19 October 1916
Crew:	26
Displacement:	Surfaced: 416 tonnes (410 tons) Submerged: 500 tonnes (492 tons)
Dimensions:	50.6m x 5.1m x 3.6m ((166ft x 17ft x 12ft)
Armament:	Three 508mm (20in) torpedo tubes, one 86mm (3.4in) gun, 18 mines
Powerplant:	Two-shaft diesel/electric motors
Surface range:	18,520km (10,000nm) at 10 knots
Performance:	Surfaced: 11.8 knots Submerged: 7.3 knots

Uebi Scebeli

Uebi Scebeli was one of the Adua class of 17 short-range vessels, only one of which, *Alagi,* was destined to survive World War II. *Uebi Scebeli* herself was an early casualty; on 29 June 1940 she was attacked by British destroyers and depth-charged to the surface, coming under fire from five of the warships. She was scuttled by her crew. The Adua class were among the best Italian submarines to be used during the war, giving good service in a variety of roles; although their surface speed was low they were strong and very manoeuvrable. The early boats of the class took part in the Spanish Civil War and all except one (the *Macalle,* which was in the Red Sea) operated in the Mediterranean. Two of the Aduas were modified in 1940-41 for the carriage of human torpedoes. These boats might well be described as the most versatile in Italian service.

Country:	Italy
Launch date:	12 January 1937
Crew:	45
Displacement:	Surfaced: 691 tonnes (680 tons) Submerged: 880 tonnes (866 tons)
Dimensions:	60.2m x 6.5m x 4.6m (197ft 6in x 21ft 4in x 15ft)
Armament:	Six 530mm (21in) torpedo tubes, one 100mm (4in) gun
Powerplant:	Two diesel engines, two electric motors
Surface range:	9260km (5000nm) at 8 knots
Performance:	Surfaced: 14 knots Submerged: 7 knots

Upholder

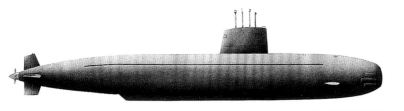

The first of a new class of diesel-electric attack submarines, HMS *Upholder* was ordered from Vickers SEL on 2 November 1983 and orders were placed for three more, named *Unseen, Ursula* and *Unicorn.* Plans for more of the class were dropped. The structure comprises a single-skinned NQ1 high tensile steel hull. The boats have an endurance of 49 days, and can remain submerged for 90 hours at 3 knots. All were based at Devonport as part of the 2nd Submarine Squadron, but by the mid-1990s all were awaiting disposal as the Royal Navy decided to have an all-nuclear submarine arm. They were sold to Canada, and renamed *Chicoutimi, Victoria, Corner Brook* and *Windsor* respectively. En route to Canada in 2004, *Chicoutimi* suffered a series of fires after seawater entered her during repairs in a storm, delaying her commissioning until 2015.

Country:	Britain
Launch date:	December 1986
Crew:	47
Displacement:	Surfaced: 2203 tonnes (2168 tons) Submerged: 2494 tonnes (2455 tons)
Dimensions:	70.3m x 7.6m x 5.5m (230ft 7in x 25ft x 17ft 7in)
Armament:	Six 533mm (21in) torpedoes; Sub Harpoon SSMs
Powerplant:	Single shaft, diesel/electric motors
Surface range:	14,816km (8000nm) at 8 knots
Performance:	Surfaced: 12 knots Submerged: 20 knots

V class

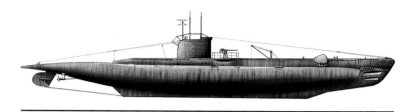

Launched in 1943-44, the Royal Navy's V-class submarines were generally similar to the U class except for their length, slightly greater because of their finer ends. In fact, the first seven boats in this category were U-class vessels but only *Upshot* and *Urtica* were actually launched, the others being cancelled. Of the 27 boats in the class, 12 went to foreign navies before the end of the war. *Variance, Venturer, Viking* and *Votary* were allocated to the Royal Norwegian Navy as the *Utsira, Utstein, Utvaer* and *Uthaug; Vineyard* and *Vortex* became the French *Doris* and *Morse; Veldt, Vengeful, Virulent* and *Volatile* went to the Royal Hellenic Navy as the *Pipinos, Delfin, Argonaftis* and *Triaina;* while *Vulpine* was assigned to the Royal Danish Navy as the *Storen. Morse,* ex-*Vortex,* also went to Denmark after her French service.

Country:	Britain
Launch date:	19 September 1944 (HMS Vagabond)
Crew:	37
Displacement:	Surfaced: 554 tonnes (545 tons)
	Submerged: 752 tonnes (740 tons)
Dimensions:	61m x 4.8m x 3.8m (200ft x 16ft x 12ft 9in)
Armament:	Four 533mm (21in) torpedo tubes; one 76mm (3in) gun
Powerplant:	Twin screws, diesel/electric motors
Surface range:	7041km (3800nm) at 8 knots
Performance:	Surfaced: 11.25 knots
	Submerged: 9 knots

Valiant

Valiant, Britain's second nuclear submarine, was slightly larger than the first, *Dreadnought*, though of basically the same design. Like *Dreadnought*, she was bult by Vickers-Armstrong. *Valiant* was originally scheduled to be completed in September 1965, but work was held up because of the priority given to the Polaris missile-armed submarines of the Resolution class, and she was not commissioned until 18 July 1966. She was followed by a sister ship, HMS *Warspite*, commissioned in April 1967, and by three Churchill-class vessels, which were modified Valiant-class SSNs and were somewhat quieter in service, having benefited from the experience gained in operating the earlier boats. *Valiant* and *Warspite*, together with the Churchills, were paid off in the late 1980s, following the full deployment of the Trafalgar class SSNs.

Country:	Britain
Launch date:	3 December 1963
Crew:	116
Displacement:	Surfaced: 4470 tonnes (4400 tons) Submerged: 4979 tonnes (4900 tons)
Dimensions:	86.9m x 10.1m x 8.2m (285ft x 33ft 3in x 27ft)
Armament:	Six 533mm (21in) torpedo tubes
Powerplant:	Nuclear, one pressurized water reactor
Range:	Unlimited
Performance:	Surfaced: 20 knots Submerged: 29 knots

Vanguard

The decision to buy the US Trident I (C4) submarine-launched ballistic-missile system was announced by the UK government on 15 July 1980. A little under two years later, the government announced that it had opted for the improved Trident II system, with the more advanced D5 missile, to be deployed in four SSBNs: these would be named *Vanguard, Victorious, Vigilant* and *Vengeance*. The four British Vanguard-class Trident submarines incorporate a missile compartment based on that of the Ohio-class boats, but with 16 rather than 24 launch tubes. Trident II D5 is able to carry up to 14 warheads of 100–120kT per missile, each having sufficient accuracy to hit underground missile silos and command bunkers, but low-yield sub-strategic warheads are also carried. The Vanguards will be replaced by the Dreadnought class currently being built.

Country:	Britain
Launch date:	4 March 1992
Crew:	135
Displacement:	Surfaced: not available Submerged: 16,155 tonnes (15,900 tons)
Dimensions:	149.9m x 12.8m x 12m (491ft 10in x 42ft x 39ft 5in)
Armament:	Sixteen Lockheed Trident D5 missiles; four 533mm (21in) torpedoes
Powerplant:	One nuclear PWR; two turbines
Range:	Unlimited
Performance:	Surfaced: not available Submerged: 25 knots

Västergotland

V*ästergotland* was the lead boat of a class of four SSKs, all named after
Swedish provinces. The others are *Hälsingland, Södermanland* and
Ostergotland. The design contract for the class was awarded to Kockums of
Malmø on 17 April 1978. The boats were optimized for operations in the Baltic,
especially in shallow coastal waters. Torpedo load comprises 12 FFV Type 613
and six FFV Type 431/450 wire-guided weapons, effective to a range of 20km
(10.8nm) at a speed of 45 knots. All four were refitted with an air-independent
propulsion system; *Södermanland* and *Ostergotland* were recommissioned in
2003–4, while the other two were sold to the Singaporean Navy as the Archer
class, adapted for use in tropical waters. *Ostergotland* was decommissioned in
2021. *Södermanland* was refitted again in 2022 to keep her in service until 2028.

Country:	Sweden
Launch date:	17 September 1986
Crew:	28
Displacement:	Surfaced: 1087 tonnes (1070 tons) Submerged: 1161 tonnes (1143 tons)
Dimensions:	48.5m x 6.1m x 5.6m (159ft 1in x 20ft x 18ft 5in)
Armament:	Six 533mm (21in)and three 400mm (15.75in) torpedo tubes
Powerplant:	Single shaft, diesel/electric motors
Surface range:	Not known
Performance:	Surfaced:11 knots Submerged: 20 knots

Velella

In 1931 Portugal ordered two submarines from the CRDA Yards, but later cancelled them for economic reasons, so they were completed for the Italian Navy under the names *Argo* and *Velella*. This accounted for the lengthy delay between the boats being laid down in October 1931 and their respective launch dates of November and December 1936; *Argo* was completed on 31 August 1937 and *Velella* the next day. *Velella* saw much war service, sinking at least two merchant vessels, before being sunk herself by HM submarine *Shakespeare* (Lt Ainslie) in the Gulf of Salerno. Her sister vessel *Argo* was scuttled in the Monfalcone CRDA yards on 11 September 1943 to avoid being captured by the Germans after the Armistice. The Germans were denied the use of many potentially valuable warships in this way.

Country:	Italy
Launch date:	12 December 1936
Crew:	46
Displacement:	Surfaced: 807 tonnes (794 tons)
	Submerged: 1034 tonnes (1018 tons)
Dimensions:	63m x 6.9m x 4.5m (207ft x 22ft 9in x 14ft 8in)
Armament:	Six 533mm (21in) torpedo tubes; one 100mm (3.9in) gun
Powerplant:	Twin screws, diesel/electric motors
Surface range:	9260km (5000nm) at 8 knots
Performance:	Surfaced: 14 knots
	Submerged: 8 knots

Victor III

An improvement on the Victor II, the first Victor III (Project 671RTM/RTMK *Shchuka*) class SSN was completed at Komsomolsk in 1978, and production proceeded at a rapid rate at both this yard and Leningrad up to the end of 1984, after which it decreased. The Victor III incorporated major advances in acoustic quietening, so that the vessel's radiated noise level was about the same as that of the American Los Angeles class. Apart from anti-ship and ASW submarines and torpedoes, the Victor IIIs were armed with the SS-N-21 Samson submarine-launched cruise missile, which had a range of 3000km (1620nm) at about 0.7M and carried a 200kT nuclear warhead. A total of 26 Victor III class boats were built, with all but two decommissioned by 2024. Its design successor was the Akula (Project 971 *Shchuka-B*) class, the first of which was launched in 1984.

Country:	Russia
Launch date:	1978
Crew:	100
Displacement:	Surfaced: not available Submerged: 6400 tonnes (6300 tons)
Dimensions:	104m x 10m x 7m (347ft 9in x 32ft 10in x 23ft)
Armament:	Six 533mm (21in) torpedo tubes; SS-N-15/16/21 SSMs
Powerplant:	Single screw, nuclear PWR, turbines
Range:	Unlimited
Performance:	Surfaced: 24 knots Submerged: 30 knots

Virginia

Built as a cheaper alternative to the Seawolf class, the Virginia class are designed for a range of missions, including anti-submarine, anti-ship, land attack and special operations support. They are considered one of the most advanced submarine classes in the world. They have been designed to use 'off the shelf' equipment and modular construction to reduce costs. They are the first submarines to use photonic sensors rather than a traditional periscope. As of 2024, 38 Virginia class have been ordered, with 25 completed. Under the AUKUS agreement, the Australian Navy will receive at least three Virginia class boats in the 2030s before the AUKUS class enters service. The US Navy is currently designing an Improved Virginia/SSN(X) class intended to replace both the Virginia and Seawolf SSN classes.

Country:	USA
Launch date:	16 August 2003
Crew:	135
Displacement:	Surfaced: N/A tonnes (715 tons)
	Submerged: 7900 tonnes (7800 tons)
Dimensions:	115m x 10m x 9.7m (377ft x 33ft x 32ft)
Armament:	Twelve vertical BGM-109 Tomahawk SLCM tubes; four 53mm (21in) torpedo tubes; 38 weapons
Powerplant:	One S9G PWR, 40,000hp; two steam turbines; pump jet propulsor
Surface range:	Unlimited
Performance:	Surfaced: N/A
	Submerged: 25 knots

W2

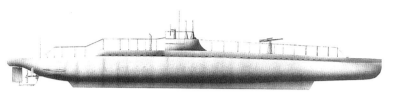

Following a fact-finding visit to Fiat-San Giorgio in 1911, a British Admiralty team next went to the Schneider Yard at Toulon to study submarine designs using the then revolutionary and advanced double-hull method of submarine construction, and also drop collars for torpedoes, enabling the weapons to be attached externally. As a result of these investigations, the Admiralty placed an order with Armstrong-Whitworth for four new boats using these innovations and known as the W class, the first two being laid down in 1913. By 1916, however, the Royal Navy had a surplus of non-standard medium-sized submarines, and so it was decided to hand *W1* and *W2* over to the Italian Navy. They had poor manoeuvrability and often suffered diesel breakdowns. Both saw limited war service, and were used mainly for training. *W2* was stricken in 1919.

Country:	Italy
Launch date:	February 1915
Crew:	19
Displacement:	Surfaced: 336 tonnes (331 tons)
	Submerged: 507 tonnes (499 tons)
Dimensions:	52.4m x 4.7m x 2.7m (172ft x 15ft 5in x 8ft 10in)
Armament:	Two 457mm (18in) torpedo tubes
Powerplant:	Twin screws, diesel/electric motors
Surface range:	4630km (2500nm) at 9 knots
Performance:	Surfaced: 13 knots
	Submerged: 8.5 knots

Walrus

Walrus was one of an eight-unit class developed from earlier American submarines. Trouble was experienced at first with her NLSE diesel engines, and in spite of these being throroughly stripped and overhauled the problem was never completely cured in three of the class. The diesel engines developed 950hp and the electric motors 680hp. Diving depth was 61m (200ft). *Walrus,* the last US submarine to be given a name for many years to come, was later renumbered *K4.* She served in the Azores during World War I, and was broken up in 1931. Many of the American submarine classes built during the World War I period were designed for purely defensive use, and were not really suitable for oceanic operations. Nevertheless, many were still on the active list when American entered the next war.

Country:	USA
Launch date:	March 1914
Crew:	31
Displacement:	Surfaced: 398 tonnes (392 tons)
	Submerged: 530 tonnes (521 tons)
Dimensions:	47m x 5m x 4m (153ft 10in x 16ft 9in x 13ft 2in)
Armament:	Four 457mm (18in) torpedo tubes
Powerplant:	Twin screws, diesel/electric motors
Surface range:	8334km (4500nm) at 10 knots
Performance:	Surfaced: 14 knots
	Submerged: 10.5 knots

Walrus

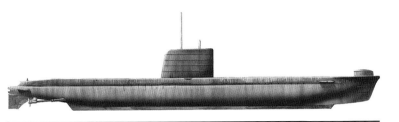

HM submarine *Walrus* was one of the eight-strong Porpoise class; these were the first operational British submarines designed after the war to be accepted into service. They were large boats with a good radius of action, which meant that they were able to undertake continuous submerged patrols in any part of the world. The design stress, in fact, was on endurance, both on the surface and submerged, whether on batteries or 'snorting'. The snorkel equipment was designed to give maximum charging facilities and to operate in rough sea conditions. Both air and surface warning radar could be operated at periscope depth as well as on the surface. The boats in the class were *Cachalot, Finwhale, Grampus, Narwhal, Porpoise, Rorqual, Sealion* and *Walrus;* the latter continued in service until the early 1980s.

Country:	Britain
Launch date:	22 September 1959
Crew:	71
Displacement:	Surfaced: 2062 tonnes (2030 tons) Submerged: 2444 tonnes (2405 tons)
Dimensions:	73.5m x 8.1m x 5.5m (241ft x 26ft 6in x 18ft)
Armament:	Eight 533mm (21in) torpedo tubes
Powerplant:	Two shafts, diesel/electric motors
Surface range:	16,677km (9000nm) at 10 knots
Performance:	Surfaced: 12 knots Submerged: 17 knots

Walrus

In 1972 the Royal Netherlands Navy identified a need for a new class of submarine to replace the elderly Dolfijn and Potvis classes. The new design evolved as the Walrus class, and was based on the *Zwaardvis* hull form with similar dimensions and silhouette, but with more automation, a smaller crew, more modern electronics, X-configuration control surfaces and the French MAREI high-tensile steel hull material that permits a 50 per cent increase in maximum diving depth to 300m (985ft). The first unit, *Walrus,* was laid down in 1979 for commissioning in 1986, but in August that year she suffered a serious fire while in the final stage of completion, so she was not completed until 1991. Despite the intensity of the blaze, her hull had luckily escaped serious damage. *Walrus* was decommissioned in 2023, but the other three boats remain active.

Country:	Netherlands
Launch date:	28 October 1985
Crew:	49
Displacement:	Surfaced: 2490 tonnes (2450 tons) Submerged: 2800 tonnes (2755 tons)
Dimensions:	67.5m x 8.4m x 6.6m (222ft x 27ft 7in x 21ft 8in)
Armament:	Four 533mm (21in) torpedo tubes
Powerplant:	Single screw, diesel/electric motors
Surface range:	18,520km (10,000nm) at 9 knots
Performance:	Surfaced: 13 knots Submerged: 20 knots

Warspite

The bearer of a famous name in British naval history, HMS *Warspite* was the third of the Royal Navy's nuclear attack submarines to be ordered from Vickers-Armstrong; she and her predecessor *Valiant* were slightly larger than *Dreadnought*. Work on both SSNs was held up because of the urgent need to bring the Polaris missile-armed submarines of the Resolution class into service to replace the RAF's V-bombers in the strategic QRA role, and *Warspite* was not completed until April 1967. She was followed by three Churchill-class vessels, which were modified Valiant-class SSNs and were somewhat quieter in service, having benefited from the experience gained in operating the earlier boats. *Valiant* and *Warspite,* together with the Churchills, were paid off in the late 1980s, following the full deployment of the Trafalgar class SSNs.

Country:	Britain
Launch date:	25 September 1965
Crew:	116
Displacement:	Surfaced: 4470 tonnes (4400 tons)
	Submerged: 4979 tonnes (4900 tons)
Dimensions:	86.9m x 10.1m x 8.2m (285ft x 33ft 3in x 27ft)
Armament:	Six 533mm (21in) torpedo tubes
Powerplant:	Nuclear, one pressurized water reactor
Range:	Unlimited
Performance:	Surfaced: 20 knots
	Submerged: 29 knots

Whiskey

The Soviet Union's first post-war submarine, and essentially a modified version of the German Type XXI design, the Russians mass-produced 236 Whiskey class diesel-electric submarines between 1949 and 1957, using prefabricated sections. All the early variants (Whiskey I-IV) were eventually converted to the Whiskey-V configuration, with no gun armament and a streamlined sail. Some were configured for special duties operations, fitted with a deck-mounted lockout diving chamber for use by Special Forces' combat swimmers. A submarine of this class (No137) went aground inside Swedish territorial waters on 27 October 1981, near Karlskrona naval base, providing evidence that the Whiskey boats were routinely engaged in clandestine activities. Some 45 Whiskey boats were transferred to countries friendly to the Soviet Union.

Country:	Russia
Launch date:	1949 (first unit)
Crew:	50
Displacement:	Surfaced: 1066 tonnes (1050 tons) Submerged: 1371 tonnes (1350 tons)
Dimensions:	76m x 6.5m x 5m (249ft 4in x 21ft 4in x 16ft)
Armament:	Four 533mm (21in) and two 406mm (16in) torpedo tubes
Powerplant:	Twin screws, diesel/electric motors
Surface range:	15,890km (8580nm) at 10 knots
Performance:	Surfaced: 18 knots Submerged: 14 knots

X1

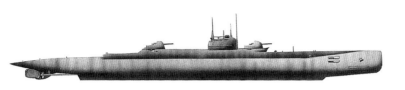

X1 was designed to evaluate the performance of a very large submarine underwater; she would probably never have existed at all had it not been for the legacy of the large German 'submarine cruisers' of World War I which, although few became operational, left behind an inflated reputation. They seemed to validate the concept of a big submersible which, carrying a heavy armament, could fight it out on the surface with destroyers and armed merchant cruisers. *X1*, unlike other prototype craft of her type, proved to have excellent handling qualities and was a steady gun platform; she was also one of the first submarines to have ASDIC detection equipment. She was the only Royal Navy vessel laid down after World War I to be scrapped before the start of the second conflict, being broken up in 1936.

Country:	Britain
Launch date:	1925
Crew:	75
Displacement:	Surfaced: 3098 tonnes (3050 tons)
	Submerged: 3657 tonnes (3600 tons)
Dimensions:	110.8m x 9m x 4.8m (363ft 6in x 29ft 10in x 15ft 9in)
Armament:	Six 533mm (21in) torpedo tubes; four 132mm (5.2in) guns
Powerplant:	Twin screws, diesel/electric motors
Surface range:	Not known
Performance:	Surfaced: 20 knots
	Submerged: 9 knots

X1

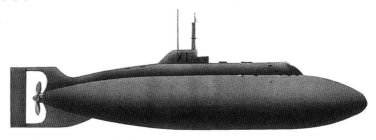

An experimental vessel, *X1* was intended to be the prototype of a series of midget submarines capable of penetrating the defences of enemy harbours, and her design was based on that of the British *X5*. *X1* normally carried a four-man crew, but on short missions she could accommodate six. She was originally fitted with a hydrogen-peroxide propulsion unit, which allowed her diesel engines to be used while submerged, and a small electric motor was fitted to allow her to 'creep' silently under water. In 1960 a hydrogen peroxide explosion blew off her bow section; the remainder of the boat remained intact. She was laid up in 1960 after repair, and later used for research purposes until 1973. *X1* was the only midget submarine built for the US Navy, which had never identified a need for such craft.

Country:	USA
Launch date:	7 September 1955
Crew:	4-6
Displacement:	Surfaced: 31 tonnes/tons Submerged: 36 tonnes/tons
Dimensions:	15m x 2m x 2m (49ft 3in x 7ft x 7ft)
Armament:	None
Powerplant:	Single screw, diesel/electric motors
Surface range:	Over 500nm
Performance:	Surfaced: 15 knots Submerged: 12 knots

X2

X *2* was a single-hulled minelaying submarine with 'saddle' tanks, based on the Austrian *U24* (ex-German *UC12*) which sank off Taranto after hitting one of her own mines. She was later raised by the Italians, who commissioned her as the *X1* and scrapped her in May 1919. *X2* was laid down on 22 August 1916 and completed on 1 February 1918. She could dive to a maximum depth of 40m (130ft); submerged range at 3 knots was 112km (60nm). A third boat in this minelayer class was also built and designated *X3*; she was launched on 29 December 1917 and completed on 27 August 1918. Both ships had nine tubes for a total capacity of 18 mines. The boats were slow and had poor manoeuvrability. They were laid up on 16 September 1940. In both world wars, minelaying was a key activity of the Italian Navy.

Country:	Italy
Launch date:	25 April 1917
Crew:	14
Displacement:	Surfaced: 409 tonnes (403 tons) Submerged: 475 tonnes (468 tons)
Dimensions:	42.6m x 5.5m x 3m (139ft 9in x 18ft x 10ft 4in)
Armament:	Two 450mm (17.7in) torpedo tubes; one 76mm (3in) gun
Powerplant:	Twin screws, diesel/electric motors
Surface range:	2280km (1229nm) at 16 knots
Performance:	Surfaced: 8.2 knots Submerged: 6.2 knots

X2

The *X2* was formerly the Italian Archimede class submarine *Galileo Galilei*. When Italy entered World War II she was based in the Red Sea, and on 19 October 1940 she was captured after a fierce battle with the British armed trawler *Moonstone* during which nearly all her officers were killed and the remaining crew, still inside the boat, were poisoned by emissions from the air conditioning system. In British service the *X2* carried the pennant number *P711*; she served in the East Indies from 1941 to 1944 as a training boat before returning to the Mediterrranean in 1944. She was scrapped in 1946. Two other Italian boats captured on active service, the *Perla* and *Tosi* (P712 and P714) were also used in the training role, the former serving in the Royal Hellenic Navy as the *Matrozos*.

Country:	Britain
Launch date:	19 March 1934
Crew:	49
Displacement:	Surfaced: 1000 tonnes (985 tons) Submerged: 1280 tonnes (1259 tons)
Dimensions:	70.5m x 6.8m x 4m (231ft 3in x 22ft 4in x 13ft 6in)
Armament:	Eight 533mm (21in) torpedo tubes; two 100mm (3.9in) guns
Powerplant:	Twin screws, diesel/electric motors
Surface range:	6270km (3379nm) at 16 knots
Performance:	Surfaced: 17 knots Submerged: 8.5 knots

X5

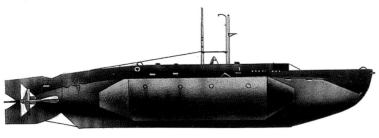

No provision was made before World War II for 'midget' submarines in the Royal Navy, and their design and production was a purely wartime expedient. Two prototypes, *X3* and *X4*, were built, and from these an operational X-type (including *X5*) was developed. The most notable event involving the class was the unsuccessful attempt to sink the German battleship *Tirpitz*. A number of X-craft were towed to Altenfjord, in northern Norway, where German surface units lay at anchor. Having successfully negotiated the minefields and barrages protecting the German ships, *X6* and *X7* managed to lay charges that damaged the battleship and put her out of action, but X5 disappeared without trace during the mission. X-craft were also used successfully against Japanese shipping in Singapore in 1945.

Country:	Britain
Launch date:	1942
Crew:	4
Displacement:	Surfaced: 27 tonnes/tons Submerged: 30 tons (29.5 tons)
Dimensions:	15.7m x 1.8m x 2.6m (51ft 6in x 6ft x 8ft 6in)
Armament:	Explosive charges
Powerplant:	Single screw, diesel/electric motors
Surface range:	Not recorded
Performance:	Surfaced: 6.5 knots Submerged: 5 knots

Xia

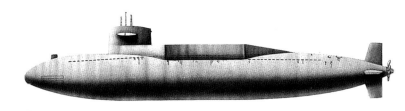

The first type of SSBN built by the People's Republic of China, the Xia (Type 092) class is roughly similar to the Russian Yankee II missile boat. The first launch of a JL-1 SLBM took place on 30 April 1982 from a submerged position near Hulodao in the Yellow Sea. The second was launched on 12 October 1982 from a specially-modified Golf-class trials submarine, and the first launch from *Changzheng 6* herself was made in 1985 and was unsuccessful, delaying the submarine's entry into service while modifications were carried out. A satisfactory launch was finally made on 27 September 1988. The JL-1 missile has a range of 972nm (1800km) and carries a single 350kT warhead. It was rumoured that a second boat was launched in 1982, and was lost in an accident in 1985. *Changzheng 6* remains on the active list but primarily as a test vessel.

Country:	China
Launch date:	30 April 1981
Crew:	140
Displacement:	Surfaced: Not known Submerged: 6604 tonnes (6500 tons)
Dimensions:	120m x 10m x 8m (393ft 7in x 33ft x 26ft 2in)
Armament:	Six 533mm (21in) torpedo tubes; 12 JL1 SLBMs
Powerplant:	Single shaft, nuclear PWR, turbo-electric drive
Range:	Unlimited
Performance:	Surfaced: Not known Submerged: 22 knots

Yankee

During the Cold War period three or four Yankee boats were on station at any one time off the eastern seaboard of the USA, with a further unit either on transit to or from a patrol area. The forward-deployed Yankees were assigned the wartime role of destroying targets such as SAC bomber alert bases and carriers/SSBNs in port, and of disrupting the American higher command echelons to ease the task of follow up ICBM strikes. As they progressively retired from their SSBN role, some Yankees were converted to carry cruise or anti-ship missiles as SSNs. Despite the removal of the ballistic missile section the overall length of the Yankee's hull was increased by 12m (39.4ft) with the insertion of a 'notch waisted' central section, housing three tubes amidships on either side, and the magazine holds up to 35 SS-N-21s or additional torpedoes and mines.

Country:	Russia
Launch date:	1967
Crew:	120
Displacement:	Surfaced: 7925 tonnes (7800 tons) Submerged: 9450 tonnes (9300 tons)
Dimensions:	129.5m x 11.6m x 7.8m (424ft 10in x 38ft x 25ft 7in)
Armament:	Six 533mm (21in) torpedo tubes; 16 SS-N-6 missiles
Powerplant:	Twin screws, two nuclear PWRs, turbines
Range:	Unlimited
Performance:	Surfaced: 20 knots Submerged: 30 knots

Yuushio

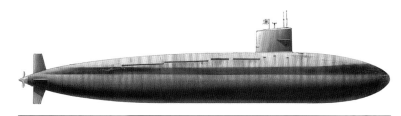

Essentially a development of the earlier Uzushio class with an increased diving depth capability, the first of the Yuushio diesel-electric submarines was laid down in December 1976 and completed in 1980. Of double-hull construction, the boats follow the US Navy practice of having a bow sonar array with the torpedo tubes moved to an amidships position and angled outwards. The names of the class after *Yuushio* are *Mochishio, Setoshio, Okishio, Nadashio, Hamashio, Akishio, Takeshio, Yukishio* and *Sachishio*. The class is equipped to fire the Sub Harpoon SSM and is fitted with indigenously-designed ASW and anti-ship torpedoes. All boats of the class were still operational in the late 1990s. The Japanese Maritime Self-Defence Force maintains some 18 patrol submarines on operational status at any one time.

Country:	Japan
Launch date:	29 March 1979
Crew:	75
Displacement:	Surfaced: 2235 tonnes (2200 tons)
	Submerged: 2774 tonnes (2730 tons)
Dimensions:	76m x 9.9m x 7.5m (249ft 3in x 32ft 6in x 24ft 7in)
Armament:	Six 533mm (21in) torpedo tubes
Powerplant:	Single shaft, diesel/electric motors
Surface range:	17,603km (9500nm) at 10 knots
Performance:	Surfaced: 12 knots
	Submerged: 20 knots

Zeeleeuw

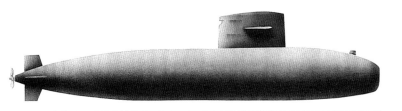

The *Zeeleeuw* (Sealion) is a Walrus-class SSK, a design based on the *Zwaardvis* hull form with similar dimensions and silhouette, but with more automation, a smaller crew, more modern electronics, X-configuration control surfaces and the French MAREI high-tensile steel hull material that permits a 50 per cent increase in maximum diving depth to 300m (985ft). The first unit, *Walrus,* was laid down in 1979 for commissioning in 1986, but in August that year she suffered a serious fire that destroyed her wiring and computers while in the final stage of completion, so she was not commissioned until March 1992, two years after *Zeeleeuw.* The other two vessels in the class are *Dolfijn* and *Bruinvis.* Two of the class are being upgraded to remain operational until the new four-boat Orka class enters service from the mid-2030s onwards.

Country:	Netherlands
Launch date:	20 June 1987
Crew:	49
Displacement:	Surfaced: 2490 tonnes (2450 tons) Submerged: 2800 tonnes (2755 tons)
Dimensions:	67.5m x 8.4m x 6.6m (222ft x 27ft 7in x 21ft 8in)
Armament:	Four 533mm (21in) torpedo tubes
Powerplant:	Single screw, diesel/electric motors
Surface range:	18,520km (10,000nm) at 9 knots
Performance:	Surfaced: 13 knots Submerged: 20 knots

Zoea

Zoea was one of three minelaying submarines built for the Italian Navy just before World War II, the others being *Atropo* and *Foca*. As first completed, their 100mm (3.9in) gun was mounted in a training turret, in the after part of the conning tower. This gun was later removed and mounted in the traditional deck position, forward of the conning tower. The class leader, *Foca*, was lost on 15 October 1940 while laying a mine barrage off Haifa, Palestine; it was thought that she had probably run into a British minefield. *Atropo* and *Zoea* survived the war and were discarded in 1947. In common with many other Italian submarines, they were in very poor condition by this time. Late in 1943, *Zoea* was used by the Allies to run supplies to British garrisons on the Aegean islands of Samos and Leros.

Country:	Italy
Launch date:	3 February 1936
Crew:	60
Displacement:	Surfaced: 1354 tonnes (1333 tons) Submerged: 1685 tonnes (1659 tons)
Dimensions:	82.8m x 7.2m x 5.3m (271ft 8in x 23ft 6in x 17ft 5in)
Armament:	Six 533mm (21in) torpedo tubes; one 100mm (3.9in) gun
Powerplant:	Twin screw diesel engines, electric motors
Surface range:	15,742km (8500nm) at 8 knots
Performance:	Surfaced: 15.2 knots Submerged: 7.4 knots

DCTN F17

The F17 was the first wire-guided torpedo to be used by the French Navy. Designed for use against surface ships from submarines, it can be employed in either the wire-guided mode or in an autonomous passive homing mode, the capability for instant switching between the two modes being provided on a control panel above the launch platform. The terminal attack phase is normally of the passive acoustic type under the torpedo's own internal control. A dual-purpose surface- or submarine-launched variant, the F17P, has been developed for the export market and has been bought by Saudi Arabia for use aboard its Madina-class frigates, and by Spain to arm its Agosta and modernized Daphne-class-submarines. The F17P differs from the F17 in having an active/ passive acoustic homing seeker, which allows autonomous operation when required.

Weapon type:	Anti-surface ship torpedo
Designer:	Direction Technique des Constructions Navales
Country of origin:	France
Weight:	1410kg (3108lb)
Dimensions:	Diameter 533mm (21in); Length 5.9m (19.4ft)
Range:	18km (9.7nm)
Warhead:	250kg (551lb) HE
Performance:	35 knots
Main operator(s):	France, Spain, Saudi Arabia

FFV Tp61 series

Designed by FFV for use against surface ship targets, the Tp61 entered service in 1967 as a non-terminal-homing wire-guided heavyweight torpedo for use by surface ships, submarines and coastal defence batteries. In 1984 the longer-range Tp613 entered service as the Tp61's successor with essentially the same propulsion system and a terminal homing seeker that utilizes an onboard computer to oversee the attack and, if necessary, to initiate previously-programmed search patterns at the target's predicted location. The torpedo's thermal propulsion system combines hydrogen peroxide with ethanol to power a 12-cylinder steam motor which produces an almost invisible wake signature. Compared with modern electrically-powered weapons at similar speed, the maximum range attainable is three to five times greater.

Weapon type:	Anti-surface vessel torpedo
Designer:	FFV
Country of origin:	Sweden
Weight:	1796kg (3959lb)
Dimensions:	Diameter 533m (21in); Length 7m (23ft)
Range:	20km (10.8nm)
Warhead:	250kg (551lb) HE
Performance:	45 knots
Main operator(s):	Sweden, Norway

Harpoon

In 1967, following the sinking of the Israeli destroyer *Eilat* by a Russian-built Styx anti-ship missile, the US Navy began to show serious interest in developing an advanced weapon of this type. The result was a formal proposal that led to the McDonnell Douglas Harpoon, a weapon that has been constantly updated over the years to match the changing threat. Harpoon, which can be launched from surface vessels, aircraft and submarines (the Sub-Harpoon version) is a highly effective weapon. Flight control is achieved by cruciform rear fins. One round is capable of destroying a guided missile boat, two will disable a frigate, four will knock out a misisile cruiser, and five will destroy a Kirov-class battlecruiser or a Kiev-class aircraft carrier. The Harpoon has been a great export success in Europe and the Middle East.

Weapon type:	Anti-ship missile
Designer:	McDonnell Douglas
Country of origin:	USA
Weight:	681kg (1498lb) with booster
Dimensions:	Diameter 343mm (13.5in); Length 4.62m (15.2ft)
Range:	160km (86nm)
Warhead:	227kg (500lb) blast-fragmentation HE
Performance:	0.85M
Main operator(s):	NATO and allied navies

M4

Twice the weight of the M20, the M4 can be fired more rapidly and from a greater operating depth than its predecessor. The missile entered service in 1985 aboard *L'Inflexible*, France's sixth ballistic-missile submarine, allowing the French Navy to maintain three vessels on patrol at all times. The design of the missile was started in 1976 and it was fired for the first time in 1980. In its original form the M4 had a range of 4,500km (2425nm), but an improved version with a range of 5000km (2695nm) entered service in 1987 on *Le Tonnant*, when that vessel completed its retrofit. The additional range was obtained by installing the lighter TN71 nuclear warhead. An advanced version of the M4, the Aérospatiale M45/TN75, armed the submarines of Le Triomphant class. This was replaced by the M51, with a range of 8000km (4300nm) in 2010.

Weapon type:	Submarine-launched ballistic missile
Designer:	Aérospatiale
Country of origin:	France
Weight:	35,073kg (77,323lb)
Dimensions:	Diameter 1.92m (6ft 3in); Length 11m (36ft 3in)
Range:	4000km (2156nm)
Warhead:	Six MIRV with TN70 150kT nuclear warheads
Performance:	Solid-propellant rocket
Main operator(s):	France

Mk37

The original Mk37 Mod 0 heavyweight torpedo entered service in 1956 as a submarine- and surface ship-launched acoustic-homing free-running torpedo. As operational experience with the weapon accumulated, many Mod 0 torpedoes were refurbished to bring them up to Mk37 Mod 3 standard. Although useful in the ASW role these free-running torpedoes, which could dive to 300m (985ft), were not suited to really long sonar detection ranges as during the weapon's run to a predicted target location it was possible that the target might perform evasive manoeuvres, taking it out of the 640m (2100ft) acquisition range of the weapon's seeker head. Successive models of the Mk37 therefore became wire-guided, the first entering service with the USN in 1962. The Mk37 was withdrawn in the 1980s.

Weapon type:	ASW torpedo
Designer:	Westinghouse
Country of origin:	USA
Weight:	767kg (1690lb)
Dimensions:	Diameter 484mm (19in); Length 3.52m (11ft 6in)
Range:	18.3km (9.8nm)
Warhead:	150kg (330lb) HE
Performance:	33.6 knots
Main operator(s):	USA and NATO

Mk46

Development of the lightweight Mk46 active/passive acoustic homing torpedo began in 1960, the first rounds of the air-launched Mk46 Mod 0 variant being delivered in 1963. The new torpedo achieved twice the range of the Mk44, which it replaced, could dive deeper (460m/1500ft against 300m/984ft) and it was 50 per cent faster (45 knots against 30). The improvement stemmed from the use of a new type of propulsion system. In the Mod 0 this was a solid-fuel motor, but as a result of maintenance difficulties it had to be changed to the Otto-fuelled thermo-chemical cam engine in the follow-on Mk46 Mod 1, which first entered service in 1967. The latest version is the Mk46 NEARTIP (NEAR-Term Improvement Program), designed to enhance the torpedo's capability against vessels with anechoic coatings.

Weapon type:	ASW torpedo
Designer:	Honeywell
Country of origin:	USA
Weight:	230kg (508lb)
Dimensions:	Diameter 324mm (12.75in); Length 2.6m (8ft 6in)
Range:	11km (6nm)
Warhead:	43kg (95lb) HE
Performance:	40/45 knots
Main operator(s):	USA, NATO and allied navies

Mk 48

The Mk48 heavyweight torpedo is the latest in a long line of 533mm (21in) submarine-launched weapons. As a long-range selectable-speed wire-guided dual-role weapon it replaced both the Mk37 series and the US Navy's only nuclear-armed torpedo, the anti-ship Mk45 ASTOR, which was fitted with a 10kT W34 warhead. Development of the Mk48 began in 1957 when feasibility studies were initiated to meet an operational requirement eventually issued in 1960. The weapon was intended as both a surface- and a submarine-launched torpedo, but the former requirement was dropped when surface-launched weapons went out of favour. The latest variant is the Mk48 Mod5 ADCAP (ADvanced CAPability) torpedo, which has a higher-powered sonar to improve target acquisition and to reduce the effect of decoys and anechoic coatings.

Weapon type:	ASW/Anti-ship torpedo
Designer:	Westinghouse
Country of origin:	USA
Weight:	1579kg (3480lb)
Dimensions:	Diameter 533mm (21in); Length 5.8m (19ft 1in)
Range:	38km (23.75 miles)
Warhead:	294.5kg (650lb) HE
Performance:	55/60 knots
Main operator(s):	USA, Australia, Canada, Netherlands

Motofides A184 and A244

T he A184 is a dual-purpose heavyweight wire-guided torpedo produced by Whitehead Motofides and deployed on submarines and surface warships of the Italian Navy. The panoramic active/passive acoustic-homing head controls the torpedo's course and depth in the final attack phase whilst the initial wire guidance uses the launch platform's own sonar sensors to guide the weapon up to the point of acoustic acquisition. Like most modern electrically-powered torpedoes the A184 is fitted with a silver-zinc battery and has dual-speed capabilities (low speed for the passive hunting phase and high speed for the terminal attack or active phase). The Mediterranean is a very difficult environment for the effective use of torpedoes, and the smaller A244 has been designed to replace the US-built Mk 44 in Italian service with that fact in mind.

Weapon type:	ASW/Anti-ship torpedo
Designer:	Whitehead Motofides
Country of origin:	Italy
Weight:	1265kg (2789lb)
Dimensions:	Diameter 533mm (21in); Length 6m (19ft 7in)
Range:	25km (13.5nm)
Warhead:	250kg (551lb) HE
Performance:	36 knots
Main operator(s)	Italy

Polaris A3

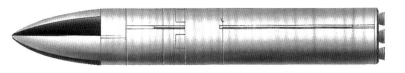

The last user of the Polaris SLBM was the Royal Navy, which had its stock of missiles re-engined in the mid-1980s so that it could provide a viable strategic missile force until the deployment of Trident. The British missiles (133 in all) were armed with three British-designed MIRVs for use against area targets such as cities and oilfields. The effect of a single high-yield warhead falls off rapidly with distance from the point of impact, whereas several lower-yield warheads around the target perimeter cause significantly more damage. The UK Polaris missiles were hardened and equipped with penetration aids as a result of a project called Chevaline, which resulted in the weapon being redesignated Polaris A3TK. The Chevaline project was prompted by Soviet developments in ABM defences.

Weapon type:	Submarine-launched ballistic missile
Designer:	Lockheed
Country of origin:	USA
Weight:	15,876kg (35,000lb)
Dimensions:	Diameter 1.4m (4ft 6in); Length 9.8m (32ft 2in)
Range:	4748km (2559nm)
Warhead:	Three 60kT MIRV, plus penetration aids and decoys
Performance:	Not known
Main operator(s):	USA, Britain

Poseidon C3

By 1964 two follow-on designs to the Polaris were under review. One subsequently evolved into the Lockheed UGM-73A Poseidon SLBM, which could use the launch tubes of exisitng SSBNs. Ultimately, 31 out of the 41 SSBNs built were refitted to carry Poseidon, although some were later fitted to carry Trident I. The Poseidon C3 entered operational service in 1970. The missile introduced the concept of multiple warheads to American SLBMs, and penetration aids were also fitted. The two-stage solid-propellant missiles were targeted mainly against soft military and industrial objectives such as airfields, storage depots and above-ground command and control facilities. One problem for an SSBN commander is that he would prefer all his SLBMs to be launched in one go, not in several groups, because each firing points to his location.

Weapon type:	Submarine-launched ballistic missile
Designer:	Lockheed
Country of origin:	USA
Weight:	29,030kg (64,000lb)
Dimensions:	Diameter 1.9m (6ft 2in); Length 10.4m (34ft)
Range:	4000-5200km (2156-2803nm) depending on warheads carried
Warhead:	Between 10 and 14 40-kT MIRV
Performance:	Not known
Main operator(s):	USA

Sea Lance

Sea Lance was approved in 1980 as the long-range anti-submarine warfare system to succeed SUBROC aboard US Navy attack submarines in the 1990s. Following target designation by the digital Mk117 fire-control system and sonars, the torpedo-tube launched missile was carried in a capsule to the surface before its single-stage solid-fuel rocket motor ignited. On clearing the surface of the sea, four small wrap-around fins at the rear of the rocket motor casing deployed automatically to stabilize the missile in flight. After booster burnout the system was jettisoned, leaving the payload to follow a ballistic trajectory to the target zone, where after deceleration the warhead would be released. The system was to have been deployed on the Los Angles class SSNs, but development was halted with the end of the Cold War.

Weapon Type:	ASW stand-off weapon
Designer:	Boeing/Gould/Hercules Aerospace
Country of origin:	USA
Weight:	1403kg (3093lb)
Dimensions:	Diameter 533mm (21in) Length 6.25m (20.5ft)
Range:	166.5km (89.7nm)
Warhead:	362.9kg (800lb) Mk50 homing torpedo
Performance:	1.5M
Operator:	USA

Spearfish

Designed to meet Naval Staff requirement 7525, the Marconi Spearfish is an advanced-capabilities wire-guided dual-role heavyweight torpedo. It was a product of the Cold War, intended to engage the new generation of high-speed deep-diving Russian submarines with the help of its HAP-Otto fuel-powered Sundstrand 21TP01 gas turbine engine with a pump-jet outlet that can drive it to speeds in excess of 60 knots. The warhead is of the directed-energy shaped-charge type, designed to penetrate Russian double-hulled submarines such as the Oscar SSGN and Typhoon SSBN. A computer enables the torpedo to make its own tactical decisions during an engagement. Work on the development prototypes began in 1982, the first in-water trials taking place the following year, and the weapon became operational in 1988.

Weapon type:	ASW/Anti-ship torpedo
Designer:	Marconi
Country of origin:	Britain
Weight:	1996kg (4400lb)
Dimensions:	Diameter 533mm (21in); Length 8.5m (27ft 11in)
Range:	36.5km (19.7nm)
Warhead:	249kg (550lb) HE
Performance:	65 knots
Main operator(s):	Britain

SS-N-6

During 1961-62 fresh impetus was given to the Soviet sea-based strategic missile programme, partly in response to the rapid deployment of the US Polaris submarine force. The Yankee class SSBNs, probably originally intended as the platform for the SS-NX-13 anti-carrier ballistic missile (tested from 1970 to 1973, but never deployed) was redirected to carry 16 SS-N-6 missiles, a design derived from the SS-11 ICBM. The SS-N-6, codenamed Sawfly by NATO, was a vast improvement over the SSN-5; it had almost twice the range, 50 per cent greater accuracy, and improved reliability. Nevertheless, its range was still limited, making it necessary for the Yankee boats (deployed from 1974) to be stationed well forward in their Atlantic and Pacific patrol areas. The deployments ceased in 1987.

Weapon type:	Submarine-launched ballistic missile
Designer:	Not known
Country of origin:	Russia
Weight:	19000kg (42,560lb)
Dimensions:	Diameter 160cm (5.4ft); Length 9.65m (31ft 7in)
Range:	3000km (1617nm)
Warhead:	One 2mT or three MIRV (Mod 2)
Performance:	Not known
Main operator(s):	Russia

SS-N-18

The SS-N-18 (NATO designation Stingray) is a fifth-generation, two-stage, liquid-fuel SLBM; it was the first Russian SLBM to have multiple warheads and was deployed in three versions, two with MIRVs and one with a single warhead. The SS-N-18 Mod 1 version was the first Soviet SLBM to feature MIRV capability. The missile was deployed on 14 Delta III class SSBNs and first became operational in 1977–78. The Mod 2 missiles carry a large single warhead with a yield of 450-1000kT. The missile has a stellar inertial guidance system with the capacity for multiple star sightings; CEP is 900m (3000ft). Each Delta III SSBN carries 16 missiles. As each of these craft was commissioned and deployed from 1976 to 1978, the Russians progressively withdrew their Yankee SSBNs from service. Russian designation of the SS-N-18 is RSM-50.

Weapon type:	Submarine-launched ballistic missile
Manufacturer:	Not known
Country of origin:	USSR
Weight:	34,000kg (75,000lb)
Dimensions:	Diameter 1.8m (5ft 11in); Length 14m (46ft)
Range:	8000km (4312nm)
Warhead:	Up to seven 200kT/500kT MIRV
Performance:	Not known
Main operator(s)	Russia

SS-N-20

The SS-N-20, allocated the NATO code-name Sturgeon, was the Soviet Union's first solid-fuel SLBM to be armed with MIRVs. A three-stage missile, it was deployed on the Soviet Navy's five operational Typhoon-class submarines and a single Golf V test submarine. Each missile was counted as carrying ten 100/200kT independently-targeted re-entry vehicles under agreed US and Soviet counting rules for the Strategic Arms Limitation Treaty (START) negotiations. The SS-N-20 is inertially guided and has an accuracy (circular error of probability/CEP) of about 600m (1800ft). Its range of 8300km (5160 miles) allows the submarine to fire the weapon from within the Arctic circle and still hit a target within continental USA. Four test flights in 1980 were failures, and were followed by two successful tests in 1981. The SS-N-20 system became operational in 1983.

Weapon type:	Submarine-launched ballistic missile
Manufacturer:	Not known
Country of origin:	USSR
Weight:	60,000kg (132,000lb)
Dimensions:	Diameter 2.2m (7ft 2in); Length 15m (49ft)
Range:	8300km (4473nm)
Warhead:	Up to ten 100-200kT MIRV
Performance:	Not known
Main operator(s):	Russia

Stingray

Designed to supplement the American Mk46 Mod 2 and to replace the Mk44 torpedoes in British service, the Stingray lightweight torpedo was the sequel to the abortive MoD in-house Mk30 and 31 programmes, cancelled in 1970. Stingray is the first British torpedo to be developed entirely by private industry and incorporates a number of technical innovations. The weapon is capable of being launched from helicopters, aircraft and surface ships over a wide range of speeds and sea states and, as a result of its unique guidance system, can be used satisfactorily in both shallow and deep waters with a high single-shot kill probability. The former was demonstrated when a Stingray dropped by a Nimrod aircraft of No 42 (TB) squadron sank the decommissioned submarine *Porpoise*, moored at periscope depth.

Weapon type:	Lightweight torpedo
Manufacturer:	Marconi
Country of origin:	Britain
Weight:	265.4kg (585.2lb)
Dimensions:	Diameter 324mm (12.75in); Length 2.6m (8ft 6in)
Range:	11.1km (6 nm)
Warhead:	40kg (88lb) shaped-charge HE
Performance:	45 knots
Main operator(s):	Britain

Thomson-Sintra Sea Mines

Thomson-Sintra produces two types of operational sea mine. The TSM5310 is an offensive ground mine fitted with a multi-sensor fusing system based on two or all of the magnetic, acoustic and pressure actuating influences, and is shaped for launching from the torpedo tube of a submarine. The sensitivity of the fusing can be adjusted before laying to suit the depth of water and the type of target likely to be encountered. The mine is armed by withdrawing two pins before it is loaded into the tube and is activated by a pre-set timing delay to allow the submarine to clear the area. The other mine, the TSM3530, is a defensive mine deployed from surface vessels fitted with mine rails. Both mines are in service with the French Navy and have been widely exported, especially to those countries that have purchased Daphne-class submarines.

Weapon type:	Sea mine
Manufacturer:	Thomson-Sintra
Country of origin:	France
Weight:	850kg (1874lb)
Dimensions:	Diameter 0.53m (1.74ft)
Range:	–
Warhead:	–
Performance:	–
Main operator(s):	France, Belgium, Malaysia, Netherlands, Pakistan, Spain

Tigerfish

The origins of the Mk24 Tigerfish heavyweight torpedo can be found as far back as 1959 in a British torpedo project codenamed 'Ongar'. By 1970 it was realized that the technology involved could not be handled solely by an in-house service approach, so the Marconi company was given the job of developing the weapon from 1972 onwards, five years after the originally envisaged in-service date. As a result of development problems the first version of the Tigerfish, the Mk24 Mod 0, entered fleet service in 1974 with less than adequate operational capability. It was only granted its full Fleet Weapon Acceptance certificate in 1979, after protracted evaluation. To rectify the problems Marconi initiated development of the Mk 24 Mod 1, but they were not entirely solved until the emergence of Mk24 Mod 2 in 1986. They were withdrawn from RN use in 2004.

Weapon type:	ASW/Anti-ship torpedo
Designer:	Marconi
Country of origin:	Britain
Weight:	1547kg (3410lb)
Dimensions:	Diameter 533m (21in); Length 6.4m (21ft 2in)
Range:	29km (15.6nm)
Warhead:	134kg (295lb) HE
Performance:	35 knots
Main operator(s):	Britain, Brazil, Turkey, Chile

BGM-109 Tomahawk

Begun in January 1974 as the US Navy's SLCM (Sea-Launched Cruise Missile) Tomahawk developed into one of the most versatile missiles in history. The multi-role Tomahawk missile is deployed in three separate naval versions: Tactical Land Attack Missile-Nuclear (TLAM-N), BGM-109B Anti-ship Tomahawk (TASM), and BGM-109C Conventional Land Attack Tomahawk (TLAM-C). Tomahawk can be launched in encapsulated form from standard USN or RN torpedo tubes, and was initially deployed on Los Angeles-class submarines in this manner. From SSN-719 *Providence* onwards, it is carried in vertical launch tubes. The first vessel to be declared operational with this installation was USS *Pittsburgh*, which entered service in November 1985. Submarine-launched Tomahawks can be fitted with combined effects bomblets for land attack missions.

Weapon Type:	Submarine-launched cruise missile
Designer:	General Dynamics
Country of origin:	USA
Weight:	1200kg (2688lb)
Dimensions:	Diameter 533mm (21in); Length 6.4m (20ft 10in)
Range:	2500km (1347nm)
Warhead:	W80 200kT nuclear, or conventional munitions
Performance:	0.7M
Operator:	USA, Britain

UGM-96A Trident I C4

The purpose of the Lockheed UGM-96A Trident I C4 missile development programme was essentially to increase the range of American SLBMs to allow the use of larger and more remote patrol areas. A three-stage solid propellant missile, Trident I was flight-tested in 1977, becoming operational two years later aboard the SSBN conversions of the Benjamin Franklin and Lafayette classes. Trident I has now been replaced by Trident II, the largest missile compatible with the launch tubes on Ohio-class SSBNs. The missile was first deployed on *Tennessee*, the ninth Ohio-class submarine, in December 1989, and a total of 312 missiles had been deployed by 1989. The missile also arms the four British Vanguard class SSBNs. The Trident II is expected to arm the forthcoming replacement US Columbia and British Dreadnought SSBN classes.

Weapon Type:	Submarine-launched ballistic missile
Designer:	Lockheed
Country of origin:	USA
Weight:	31,751kg (70,000lb)
Dimensions:	Diameter 1.89m (6ft 2in); Length 10.4m (34ft 1in)
Range:	6808km(3669nm)
Warhead:	Seven Mk4 MIRV with 100kT W-76 warheads
Performance:	Not known
Operator:	USA

Sub-Martel

Sub-Martel, properly designated Under-Sea Guided Weapon CL137, was a unilateral attempt by Britain to develop an effective submarine-launched SSM to counter the Russian N-7 system. It was to have been a collaborative effort between HSD and the French firm Matra, designers of the air-launched version, but this scheme was abandoned when the French turned their attention to development of the SM38 Exocet. The USGW was based heavily on Martel, but Matra took no part in the project. The intention was to extend the length of the Martel body and add folding flip-out wings and a booster motor. The homing head was to be developed by Marconi Space and Defence Systems. In the event, development was cancelled in 1975, after some £16 million had been spent, and orders were placed for the American Harpoon SSM instead.

Weapon Type:	Submarine-launched SSM
Designer:	Hawker Siddeley Dynamics/Matra
Country of origin:	Britain
Weight:	550kg (1213lb)
Dimensions:	Diameter 400mm (15.75in); Length 3.87m (152.4in)
Range:	30km (16nm)
Warhead:	150kg (330lb) HE
Performance:	Not known
Operator:	–

UUM-44A SUBROC

Development of the two-stage SUBROC submarine rocket began in 1958, the complete technical evaluation being finished by 1964. The first production rounds were delivered to the US Navy in the following year, and the average SSN basic load was four to six rounds. The missile was launched normally from a 533mm (21in) torpedo tube; at a safe distance from the submarine the solid-fuel rocket motor ignited and the weapon followed a short level path before pitching up out of the water. At the optimum payload release point, the missile having been steered by four jet deflectors, the 5kT yield W55 warhead was released by a combination of explosive bolts and a thrust reversal deceleration system, continuing on a ballistic trajectory to the target area. The bomb then sank to a pre-set depth before detonating.

Weapon type:	ASW rocket
Designer:	Goodyear
Country of origin:	USA
Weight:	1814kg (4000lb)
Dimensions:	Diameter 533mm (21in); Length 6.7m (22ft)
Range:	56km (30nm)
Warhead:	W55 5kT nuclear
Performance:	1.5M
Main operator(s)	USA

Index

Note: Page numbers in **bold** refer to main entries.

A class 9
A1 14
Acciaio **15**, 34
Adua class 20, 48, 68, 215, 266
Agosta **16**, 108, 220
Akula (Project 971 *Shchuka-B*) class 273
Albacore **17**
Alfa (Project 705 *Lira*) class **18**
Aluminaut **19**
Alvin 52
American Civil War 8
    H.L. Hunley 144
    Intelligent Whale 155
    Pioneer 194
American War of Independence 7
Ammiraglio Cagni class 38
Aradam **20**
Archimede (Brin class) **21**
Archimede class 33, 92, 109, 112, 282
Aréthuse 51
Argo class 93, 129
Argonaut (1897) **22**
Argonaut (1927) **23**
Arihant **24**
Ateliers Loire-Simonot, France 107, 181
Astute **25**
Astute class 25, 236
AUKUS class 25, 44, 274
Australia
    AUKUS class 44, 274
    Collins 44
    Virginia class 274

B class 9
B1 **26**
Balao class 192
Balilla **27**, 63, 75
Balilla class 27, 63, 75
ballistic missiles
    M4 **294**
    Polaris A3 **299**
    Poseidon C3 **300**
    SS-N-6 **303**
    SS-N-20 **304**
    SS-N-18 **305**
    VGM-96A Trident 1 C4 **310**
Barb 74
Barbarigo (1917) **28**
Barbarigo (1938) **29**
Barracuda class 30
Bass **30**
Battle of the Atlantic 10–11
Benjamin Franklin class 50, 310
Bernardis, Curio 55, 101, 111, 118, 178
Beta **31**
BGM-109 Tomahawk 309

Blaison **32**
Borei class 57, 217
Brazil
    Riachuelo class 215
Brin **33**
Brin class 21, 33, 110, 199
Britain
    A1 14
    Astute **25**
    B1 **26**
    C3 **36**
    C25 **37**
    Conqueror **45**
    D1 **47**
    Dreadnought **65**
    E11 **70**
    E20 **71**
    Explorer **85**
    F1 **86**
    G1 **104**
    J1 **157**
    K4 **158**
    K26 **159**
    L10 **163**
    L23 **164**
    M1 **167**
    Nautilus **174**
    Oberon (1926) **183**
    Oberon (1959) **184**
    Odin **185**
    Parthian **191**
    Porpoise **196**
    R1 **197**
    Resolution **203**
    Resurgam II **204**
    S1 **210**
    Sanguine **213**
    Swiftsure **227**
    Swordfish (1916) **228**
    Swordfish (1931) **229**
    Thames **233**
    Thistle **234**
    Torbay **236**
    Trafalgar **237**
    U class **263**
    Upholder **267**
    V class **268**
    Valiant **269**
    Vanguard **270**
    Walrus **277**
    Warspite **279**
    X1 **281**
    X2 **284**
    X5 **285**
Bronzo **34**
Brumaire class 102

C class 9, 36, 37
    C3 **36**
    C25 **37**
C1 class **35**
    C3 **36**
    C25 **37**
Cagni **38**
Calvi class 73, 121

Canada
    Chicoutimi class 267
Casabianca **39**
Casma **40**
CB12 **41**
Changzeng 1 140
Changzeng 6 286
Changzeng 11–14 242
Changzeng 20–21 242
Charlie class **42**, **43**, 190
    Charlie I class **42**
    Charlie II class **43**
Cherbourg Naval Dockyard 103, 198, 232
Chickwick 77
Chicoutimi class 267
Chile
    Carrera 215
    O'Higgins 215
China
    Han class **140**
    Romeo (Type 033) class **207**
    Type 039 (Song) class **241**
    Type 039A/041 (Yuan) class **241**
    Type 094 class **242**
    Type 096 class **242**
    Xia (Type 092) class 140, **286**
Churchill class 45, 267
Cold War
    Entemedor 77
    Evangelista Torricelli 84
    Nacken 171
    Shark 218
    Spearfish **302**
    Yankee class **287**
Collins **44**
Confederate States of America
    H.L. Hunley **144**
    Pioneer **194**
Conqueror **45**
Corallo **46**

D class 47
D1 **47**
Dagabur **48**
Dakar class 62
Dandolo **49**
Daniel Boone **50**
Daphné **51**
Daphné class 51, 59, 64, 169, 307
DCTN F17 **291**
Deep Quest **52**
Deepstar 4000 **53**
Delfino (1890) **54**
Delfino (1930) **55**
Delta I (*Murena*) class **56**
Delta II class **57**
Delta III (*Kalmar*) class **57**, 305
Delta IV (*Delfin*) class **57**
Denmark
    Dykkeren **69**
Deutschland **58**
Diablo **59**
Diaspro **60**
Dolfijn **61**, 278
Dolphin **62**
Dolphin classes 62

Domenico Millelire **63**
Doris **64**
Dreadnought **65**
Drum **66**
Dupuy de Lôme **67**, 134
Durbo **68**
Dykkeren **69**

E class 47, 70, 71, 95
E11 **70**
E20 **71**
Echo class **72**
Electric Boat Company 136, 168
Enrico Tazzoli (1935) **73**
Enrico Tazzoli (1942) **74**
Enrico Toti (1928) **75**, 191
Enrico Toti (1967) **76**
Entemedor **77**
Ersh class *(SHCH 303)* **78**
Espadon (1901) **79**
Espadon (1926) **80**
Ethan Allen class 114, 153
Ettore Fieramosca **81**
Euler **82**
Eurydice **83**
Evangelista Torricelli **84**
Explorer **85**

F1 (Britain) **86**
F1 (Italy) **87**
F4 **88**
Faa di Bruno **89**
Falklands War 13
    Conqueror 45
    Oberon class 184
Farfadet **90**
Fenian Ram **91**
Ferraris **92**
Ferro **93**
FFV Tp61 series **292**
Fiat-San Giorgio, Italy 69, 147, 275
Filippo Corridoni **94**
Fisalia **95**
Flutto **96**
Flutto class 93, 96, 129
Foca (1908) **97**
Foca (1937) **98**
Foca (1937) class 25, 98, 290
Fornier, Father 7
Foxtrot class **99**, 231
France
    Agosta **16**
    Blaison **32**
    Casabianca **39**
    Daphné **51**
    Deepstar 4000 **53**
    Doris **64**
    Dupuy de Lôme **67**
    Espadon (1901) **79**
    Espadon (1926) **80**
    Euler **82**
    Eurydice **83**
    Farfadet **90**
    Frimaire **102**
    Fulton **103**
    Galathée **107**

Goubet I **125**
Goubet II **126**
Gustave Zédé (1893) **132**
Gustave Zédé (1913) **133**
Gymnôte **134**
Henri Poincaré **143**
Nymphe **181**
Redoubtable (1967) **198**
Requin class (1924) **201**
Requin (1955) **202**
Roland Morillot **206**
Rubis (1931) **208**
Rubis (1979) **209**
Suffren class **209**
Surcouf **226**
Le Terrible **232**
Le Triomphant **238**
Francesco Rismondo **100**
Fratelli Bandiera **101**
Frimaire **102**
Fulton **103**
Fulton, Robert 7–8, 103, 173

G class **104**
Gal **105**
Galatea **106**
Galathée **107**
Galerna **108**
Galilei **109**
Galvani **110**
Gato class 59, 66, 74, 77, 84, 130, 195
Gemma **111**
General Mola **112**
George Washington **113**
George Washington class **113**
George Washington Carver **114**
Georgia **115**
Germany
    Deutschland **58**
    Type 212A class **243**
    U1 **245**
    U2 **246**
    U3 **247**
    U12 **248**
    U21 **249**
    U28 **250**
    U32 **251**
    U47 **252**
    U106 **253**
    U112 **254**
    U139 **255**
    U140 **256**
    U151 **257**
    U160 **258**
    U1081 **259**
    U2326 **260**
    U2501 **261**
    U2511 **262**
    UB4 **264**
    UC74 **265**
Ghazi **59**
Giacinto Pullino **116**
Giacomo Nani **117**
Gianfranco Gazzana Priaroggia 192
Giovanni Bausan **118**
Giovanni de Procida **119**

Giuliano Prini **120**
Giuseppe Finzi **121**
Glauco (1905) **122**
Glauco (1935) **123**
Glauco (1935) class 89, **123**
Golf I class **124**
Golf II class **124**
Gorki shipyard, Russia 42, 43, 160, 219, 231
Goubet I **125**
Goubet II **126**
Grayback **127**
Grayling **128**
Grongo **129**
Grouper **130**
Greece
    Nordenfelt I **179**
Guadalcanal 66
Guglielmo Marconi **131**
Gulf War
    Los Angeles **165**
Gustave Zédé (1893) **132**
Gustave Zédé (1913) **133**
Gymnôte **134**

H class (British) **135**
H class (Italian) **135**
H class (Russian) **136**
H1 **135**
H4 **136**
Ha 201 class **137**
Hai Lung **138**
Hajen **139**
Han class **140**
Harpoon **293**
Harushio (1967) **141**
Harushio (1989) class **142**, 189
Henri Poincaré **143**
H.L Hunley 8
HL Hunley **144**
Holland, John Phillip 8–9, 22, 91, 103, 145
Holland No 1 **145**
Holland VI **146**
Hotel class 12
Hvalen **147**

I7 **148**
I15 **149**
I15 class **149**
I21 **150**
I201 class **151**
I351 **152**
I400 class **153**
India
    Arihant **24**
    Chakra 42
    Kalvari class 215
India **154**
India class **154**
Indo-Pakistan War
    Daphné class 64, 169
    Ghazi (ex *Diablo*) 59
Intelligent Whale **155**
Isaac Peral **156**
Israel
    Dakar class 62
    Dolphin **62**

Dolphin 2 class 62
Drakon 62
Gal 105
Rahav (ex Sanguine) 213
Type 206 class 105
Type 209 class 62
Italy
  Acciaio **15**
  Aradam **20**
  Archimede **21**
  Balilla **27**
  Barbarigo (1917) **28**
  Barbarigo (1938) **29**
  Beta **31**
  Brin **33**
  Bronzo **34**
  CB12 **41**
  Cagni **38**
  Corallo **46**
  Dagabur **48**
  Dandolo **49**
  Delfino (1890) **54**
  Delfino (1930) **55**
  Diaspro **60**
  Domenico Millelire **63**
  Durbo **68**
  Enrico Tazzoli (1935) **73**
  Enrico Tazzoli (1942) **74**
  Enrico Toti (1928) **75**
  Enrico Toti (1967) **76**
  Ettore Fieramosca **81**
  Evangelista Torricelli **84**
  F1 **87**
  Faa di Bruno **89**
  Ferraris **92**
  Ferro **93**
  Filippo Corridoni **94**
  Fisalia **95**
  Flutto **96**
  Foca (1908) **97**
  Foca (1937) **98**
  Francesco Rismondo **100**
  Fratelli Bandiera **101**
  Galatea **106**
  Galilei **109**
  Galvani **110**
  Gemma **111**
  Giacinto Pullino **116**
  Giacomo Nani **117**
  Gianfranco Gazzana Priaroggia **192**
  Giovanni Bausan **118**
  Giovanni de Procida **119**
  Giuliano Prini **120**
  Giuseppe Finzi **121**
  Glauco (1905) **122**
  Glauco (1935) **123**
  Grongo **129**
  Guglielmo Marconi **131**
  H1 **135**
  Luigi Settembrini **166**
  Nazario Sauro **177**
  Nereide **178**
  Pietro Micca **193**
  Reginaldo Giuliani **199**
  Remo **200**
  Type 212A class **120**

Uebi Scebeli **266**
Velella **272**
W2 **275**
X2 **283**
Zoea **290**

J class 157
J1 **157**
Japan
  C1 class **35**
  Ha 201 class 137
  Harushio (1967) **141**
  Harushio (1989) class **142**
  I7 **148**
  I15 class **149**
  I16 35
  I21 **150**
  I201 class **151**
  I351 **152**
  I400 class **153**
  Oyashio **189**
  RO105 **205**
  Soryu class 189
  Yuushio class **288**
Jiangnan (Shanghai) shipyard 207

K class
  K4 **158**
  K26 **159**
Kalmar class 57
Kalvari class 215
Karp class 245
Kawasaki shipyard, Kobe 150, 189, 205
Kilo class **160**
Kilo II class 160
Komsomolsk, Russia 72, 160, 273
Kursk (Oscar class) 188

L3 (Russia) **161**
L3 (USA) **162**
L class
  L10 **163**
  L23 **164**
La Spezia Naval Dockyard, Italy 54, 116
Lafayette class 50, 114, 310
Lake, Simon 22
Laurenti 122, 228
Lee, Ezra 7, 240
Liuzzi class 199
Lizardfish 84
Lorient, France 143
Los Angeles **165**
Los Angeles class 165, 212, 216, 218, 301
Luigi Settembrini **166**

M class 167
M1 **167**
M4 **294**
Mackerel **168**
Mameli class 119
Marcello class 49
Marconi class 131
Marlin **168**
Marsopa **169**
McDonnell Douglas Harpoon **293**
Medusa class 87, 95

Mersenne, Father 7
mines
    Thomson-Sintra sea mines **307**
missiles *see also* ballistic missiles
    BGM-109 Tomahawk **309**
    Harpoon **293**
    Sea Lance **301**
    Sub-Martel **311**
    UUM-44A SUBROC **312**
Mitsubishi Company 189
Mk37 **295**
Mk46 **296**
Mk48 **297**
Morocco Flotilla 67
Motofides A184 & A244 **298**
Murena class 56

N class 170
N1 **170**
Näcken **171**
Narval class 51, 64, 202
Narwhal **172**
Narwhal class 172
Nautilus (1800) 8, 103, **173**, 176
Nautilus (1914) **174**, 228
Nautilus (1930) **175**
Nautilus (1954) **176**
Nazario Sauro **177**
Nereide **178**
Netherlands
    Dolfijn 61
    Orka class 289
    Walrus **278**
    Zeeleeuw **289**
Nordenfelt I **179**
November class 180
Nymphe **181**

O class (Britain) 183, 185
    Oberon (1926) **183**
    Odin **185**
O class (USA) **182**
Oberon (1926) **183**
Oberon (1959) **183**
    Odin **185**
Ohio **186**
Ohio class 50, 115, 186, 310
Oldenburg 58
Operation Landcrab
    Nautilus 175
Orka class 289
Orzel **187**
Oscar class 13, **188**, 190
    Kursk 188
Oshio class 141
Ostvenik 100
Oyashio **189**
Oyashio class 142, 189

Papa class **190**
Parthian **191**, 201
Parthian class 191
Perla class 46, 60, 111, 215
Peru
    Casma 40
Pickerel **192**

Pietro Micca **193**
Pioneer **194**
Piper **195**
Pisani class 94, 118
Poland
    Orzel **187**
Polaris A3 **299**
Porpoise **196**
Porpoise class 184, 196, 277
Portsmouth Navy Yard 30, 168, 195
Poseidon C3 **300**
Potvis class 276
Prien, Lt Cdr Gunther 252
Project 705 *Lira* **18**

R class (Britain) 197
R class (Italy) 200
R1 **197**
Rahav (ex *Sanguine*) 213
Redoubtable class (1925)
    Casabianca 39
    Henri Poincaré 143
Redoubtable class (1967)
    Redoutable 198
    Terrible 232
Reginaldo Giuliani **199**
Remo **200**
Requin class (1924) 80, 143, **201**, 202
Requin (1955) **202**
Resolution **203**
Resolution class 203, 279
Resurgam II **204**
Richson, Carl 139
Riachuelo class 215
River class 233
RO100 **205**
Roland Morillot **206**
Romeo class **207**
Rubis (1931) **208**
Rubis (1979) **209**
Rubis class 209
Russia
    Alfa (Project 705 *Lira*) class **18**
    Akula (Project 971 *Shchuka-B*) class 273
    Borei class 57, 217
    Charlie I class **42**
    Charlie II class **43**
    Delta I (*Murena*) class **56**
    Delta II class 57
    Delta III (*Kalmar*) class **57**, 305
    Delta IV (*Delfin*) class 57
    Echo class **72**
    Ersh (Pike) class **78**
    Foxtrot class **99**, 231
    Golf I **124**
    India **154**
    Kilo class 160
    L3 **161**
    November class **180**
    Oscar class 188
    Papa class **190**
    Romeo class **207**
    SHCH 303 **78**
    Severodvinsk **217**
    Sierra (Project 945) class **219**
    Tango class **231**

Typhoon class **244**
Victor III (Project 671) class **273**
Whiskey **280**
Yankee class **287**
Yasen class 217

S class (1914) 210
S class (1918) 211
S class (1931) 213, 229
S1 **210**
S28 **211**
San Francisco **212**
Sandford, Lt Richard D 36
Sanguine **213**
Santa Cruz **214**
Saphir class **208**
Sasebo Naval Yard, Japan 137
Sauro class 177
Scorpène **215**
Scotts, Britain 216, 228
Sea Lance **301**
Seawolf **216**
Seawolf class 216
Serbia, NATO strikes on 227
Severodvinsk **217**
Severodvinsk, Russia 56, 180, 188, 190, 219
Shark **218**
SHCH 303 **78**
Sierra class **219**
Singapore
    Södermanland class 271
Sirena class 106, 111
Sirène class 107, 181
Siroco **220**
Sjoormen class **221**
Skate **222**
Skate class 222
Skipjack **223**
Skipjack class 65, 218, 223
Södermanland class 271
Song (Type 039) class **241**
Soryu **224**
Spain
    Galerna **108**
    General Mola **112**
    Isaac Peral **156**
    Marsopa **169**
    Siroco **220**
Spanish Civil War
    Adua class 20, 266
    Archimede **21**
    Diaspro **60**
    Domenico Millelire **63**
    Enrico Tazzoli **73**
    Enrico Toti **75**
    Ferraris **92**
    Galatea **106**
    Galilei **109**
    Gemma **111**
    General Mola **112**
    Giovanni da Procida **119**
    Perla class 46
Spearfish **302**
Spinelli, Major Engineer 38
Squalo class 55
SS-N-6 **303**

SS-N-18 **304**
SS-N-20 **305**
Stingray **306**
Sturgeon class 172, **225**, 304
    Parche 225
Sub-Martel **311**
Suffren class **209**
Surcouf **226**
Sweden
    Hajen **139**
    Hvalen **147**
    Näcken **171**
    Södermanland class 271
    Sjoormen class **221**
    Type 471 class 44
    Västergotland **271**
Swiftsure **227**
Swiftsure class 227
Swordfish (1916) **228**
Swordfish (1931) **229**

T class 46, 234
    Thistle **234**
    Tigris 46
Tang class 225, **230**
Tango class **231**
Tench class 59, 192
Le Terrible **232**
Thames **233**
Thistle **234**
Thomson-Sintra sea mines **307**
Thresher/Permit class 225, **235**
Thyssen Nordseewerk, Germany 214
Tigerfish **308**
Tizzoni, Major Engineer 118
Todaro class 243
Tomahawk **309**
torpedoes
    DTCN F17 **291**
    FFV Tp61 series **292**
    Mk37 **295**
    Mk46 **296**
    Mk48 **297**
    Motofides A184 & A244 **298**
    Spearfish **302**
    Stingray **306**
    Tigerfish **308**
Torbay **236**
Toricelli 112
Tosi yards, Italy 119
TR1700 class 214
Trafalgar **237**
Trafalgar class 45, 236, 237, 279
    Torbay **236**
    Trafalgar **237**
Travkin, I.V 78
Tricheco 111
Trident 299, 300, **310**
Le Triomphant **238**, 294
Triomphant class 232, 238, 294
Triton **239**
Turkey
    Piri Reis (ex *Tang*) 230
    Preveze (ex *Entemedor*) 77
Turtle 7, **240**
Type 039 (Song) class **241**

Type 039A/041 (Yuan) class 241
Type 092 class 242, **286**
Type 094 class **242**
Type 096 class 242
Type 205 class 246, 247
Type 206 class 105, 247
Type 209 class 13, 41
Type 212A class 120, **243**
Type 212CD class 243
Type 471 class 44
Type 640 105
Type 2400 205
Type IX 32, 253
Type KS class 205
Type VII 32, 251, 252
Type XI 254
Type XVII 259
Type XXI 32, 151, 202, 206, 260, 261, 262, 280
Type XXIII 260, 262
Typhoon class **244**, 304

U class 55, **263**
U1 **245**
U2 **246**
U3 **247**
U12 **248**
U21 **249**
U28 **250**
U32 **251**
U42 see *Balilla*
U47 **252**
U106 **253**
U112 **254**
U123 32
U139 **255**
U140 **256**
U151 **257**
U160 **258**
U1081 **259**
U2326 **260**
U2501 **261**
U2511 **262**
UB4 **264**
UC74 **265**
Uebi Scebeli **266**
UGM-96A Trident I C4 **310**
Upholder **267**
Upholder class 267
USA
   Albacore 17
   Aluminaut 19
   Argonaut 22, **23**
   Bass **30**
   Daniel Boone **50**
   Deep Quest **52**
   Diablo **59**
   Drum **66**
   F4 **88**
   Fenian Ram **91**
   George Washington **113**
   George Washington Carver **114**
   Georgia **115**
   Grayback **127**
   Grayling **128**
   Grouper **130**
   H4 **136**

Holland No 1 **145**
Holland VI **146**
Intelligent Whale **155**
L3 **162**
Los Angeles **165**
Marlin **168**
N1 **170**
Narwhal **172**
Nautilus **173, 175, 176**
O class **182**
Ohio **186**
Pickerel **192**
Piper **195**
S28 **211**
San Francisco **212**
Seawolf **216**
Shark **218**
Skate **222**
Skipjack **223**
Sturgeon **225**
Tang **230**
Thresher/Permit class **235**
Triton **239**
Turtle **240**
Virginia class **274**
   Walrus **278**
   X1 **282**
UUM-44A SUBROC **312**
Uzushio class 142, **288**

V class (Britain) **268**
V class (USA) 175
Valiant **269**
Valiant class 45, 269
Vanguard 238, **270**
Vanguard class 203, 270
Västergotland **271**
Velella **272**
Venice Naval Yard, Italy 31, 122
Vickers, Barrow-in-Furness, Britain 9, 14, 100, 105, 174
Vickers-Armstrong 85, 233, 263, 269, 279
Vickers Shipbuilding and Engineering Ltd 237, 267
Victor class 42, 212, 231, 273
   Victor III (Project 671) class **273**
Virginia class **274**

W2 **275**
Walrus (Britain) **277**
Walrus (Netherlands) 278, **289**
Walrus (USA) **276**
Warspite **279**
weapons *see* ballistic missiles, mines, missiles, torpedoes
Westinghouse Electric Corporation 53
Whiskey class 230, **280**
Wilkins, John 7
World War I 9–10
   B1 **26**
   Balilla **27**
   Beta **31**
   C3 **36**
   C25 **37**
   D1 **47**
   E11 **70**

E20 71
Euler 82
F1 86, 87
Fisalla 95
Frimaire 102
Gl 104
Giacinto Pullino 116
Grayling 128
Gustave Zédé 133
HI 135
J1 157
K4 158
L3 162
L10 163
N1 170
Nereide 178
O class 182
R1 197
S1 210
Swordfish 228
U21 250
U139 255
U151 257
U160 258
UB4 264
UC74 265
W2 275
Walrus 276
X2 283
World War II 10–2
  Acciaio 15
  Aradam 20
  Archimede 21
  Argonaut 23
  Barbarigo 29
  Blaison 32
  Brin 33
  Bronzo 34
  Cl class 35
  Cagni 38
  Casabianca 39
  CB12 41
  Corallo 46
  Dagabur 48
  Dandolo 49
  Delfino 55
  Diaspro 60
  Drum 66
  Durbo 68
  Enrico Tazzoli 73
  Enrico Toti 75
  Ersh (SHCH 303) 78
  Espadon 80
  Ettore Fieramosca 81
  Eurydice 83
  Faa di Bruno 89
  Ferraris 92
  Filippo Corridoni 94
  Flutto 96
  Foca 98
  Francesco Rismondo 100
  Fratelli Bandiera 101
  Galatea 106
  Galathée 107
  Galilei 109
  Galvani 110

Gemma 111
Giovanni da Procida 119
Giuseppe Finzi 121
Glauco 123
Grayling 128
Grongo 129
Grouper 130
Guglielmo Marconi 131
HI 135
Henri Poincaré 143
I7 148
I15 149
I21 150
I351 152
L3 161
L23 164
Luigi Settembrini 166
Nautilus 175
Odin 185
Orzel 187
Parthian 191
Pickerel 192
Pietro Micca 193
Piper 195
Porpoise 196
Reginaldo Giuliani 199
Remo 200
Requin 201
RO100 205
Rubis 208
S28 211
Surcouf 226
Swordfish 229
Thames 233
Thistle 234
U class 263
U2 246
U32 251
U47 252
U106 253
Uebi Scebli 266
Velella 272
X2 283, 284
X5 285
Zoea 290

X1 (Britain) 281
X1 (USA) 282
X2 (Britain) 284
X2 (Italy) 283
X5 282, 285
Xia (Type 092) class 140, 286

Yankee class 56, 287, 303
Yasen class 217
Yuushio class 142, 288

Zeeleeuw 289
Zoea 290
Zwaardvis class 138